T0134648

Springer Topics in Signal Processing

Volume 14

Series editors

Jacob Benesty, Montreal, Canada
Walter Kellermann, Erlangen, Germany

More information about this series at http://www.springer.com/series/8109

Jagannath Malik · Amalendu Patnaik
M.V. Kartikeyan

Compact Antennas for High Data Rate Communication

Ultra-wideband (UWB) and Multiple-Input-Multiple-Output (MIMO) Technology

Jagannath Malik
Department of Electrical Engineering
Ulsan National Institute of Science and Technology
Ulsan
Republic of Korea

M.V. Kartikeyan
Department of Electronics and Communication Engineering
Indian Institute of Technology
Roorkee
India

Amalendu Patnaik
Department of Electronics and Communication Engineering
Indian Institute of Technology
Roorkee
India

ISSN 1866-2609 ISSN 1866-2617 (electronic)
Springer Topics in Signal Processing
ISBN 978-3-319-87491-3 ISBN 978-3-319-63175-2 (eBook)
DOI 10.1007/978-3-319-63175-2

Printed on acid-free paper

This Springer imprint is published by Springer Nature
The registered company is Springer International Publishing AG
The registered company address is: Gewerbestrasse 11, 6330 Cham, Switzerland

... To my beloved mother

M.V. Kartikeyan

Preface

Wireless communication systems have become an integral part of humans' everyday life. Wireless communication technologies enable consumers to have access to a wide range of services comfortably at any time from any place. The limitations imposed on the people's communication by geographical locations are eliminated with the introduction of wireless mobile access points. The Internet allowed people to access and share information in a much faster way. Satellite TV and radio are one of the sources for peoples entertainment. Wired devices have been replaced with cordless devices. Wireless networking technologies such as Bluetooth, wireless fidelity (Wi-fi), and wireless personal area networks (WPAN) enhanced the way portable devices stay interconnected. The demand for high speed, high capacity, and secure wireless communications is increasing day by day to keep up with the ever increasing needs of the humans.

The upper limit on the capacity or data rate attainable in an additive white Gaussian noise (AWGN) channel characterized as band-limited ideally is given by Shannon-Nyquist Criterion [1, 2]. The increment in transmission rate of data is possible either by raising the transmission power level or extending the occupied bandwidth. The channel capacity increases logarithmically as the signal-to-noise (SNR) increases. Existing wireless systems operate at fixed frequencies and bandwidth with strict regulations imposed by Federal Communication Commission (FCC). Higher speeds are attainable with raising the transmission signal power. A large number of portable devices use battery as a source of power which limits the transmitted power. Apart from that a higher transmitted power leads to the potential interference with nearby communication systems, as wireless system is interference limited. Consequently, a large spectral bandwidth is the best possible solution to achieve higher data rates. However, the spectrum is scares and comes under the strict regulations of FCC.

In order to achieve higher data rates with the limited available resources and imposed constraints, wireless communication technology needs to be pushed beyond the physical limits of the propagation channel. To address the issue, two promising technologies have been evolved in the past decade. One is the ultra-wideband (UWB) communication and the other one is multiple-input-multiple-output (MIMO) communication system. The approach to achieve higher data rate over Shannon's limit using these technology is different from one another. The former technology takes the advantage of short time-domain un-modulated pulse transmission, whereas the later takes the advantage of multipath propagation characteristics of wireless channel to achieve a higher data rate. These technologies can coexist with existing narrowband communication standards without any interruption/interference.

In February 2002, a wide 7.5 GHz spectrum beginning at 3.1 GHz and ending at 10.6 GHz was released by Federal Communications Commission (FCC) for unlicensed use permitting the implementation of ultra-wideband (UWB) technology for commercial purposes [3]. UWB technology has drawn significant attention with this new spectrum allocation. UWB technology offers a promising solution to the high data rate transmission and RF spectrum drought. Ultra-wideband communications are a game changing wireless technology with many inherent merits. In UWB technology, large amounts of digital data are transmitted over a short range over a large portion of the frequency spectrum utilizing low powered, short pulse radio signals. UWB technology is capable of delivering performance not attainable by conventional narrow band wireless technologies. Some advantages of UWB technology include high data rate are consumption of less average power, better immunity to multipath cancellation, higher range measurement accuracy, and range resolution, low cost [4]. By using very low powered, short pulse, broad spectrum radio signals, the interference caused when compared to the other narrow band systems is very less, and hence coexistence with already existing other narrowband conventional wireless technologies which include wireless local area network (WLAN), wireless fidelity (Wi-fi), and Bluetooth is possible. These features of UWB technology make it suitable for applications such as covert communication links, surveillance, Internet with very high-speed broadband accession in short range, ground-penetrating radar with high resolution, through-wall imaging, precision navigation, RF tracking and positioning, and high bit rate WPAN [5].

Antenna is a crucial part of any system because ultimately any wireless system needs to use an antenna to convert the signal to electromagnetic waves for propagation in free space and the received electromagnetic waves back to signal. The performance of UWB systems is highly dependent upon antenna characteristics. A high performance in terms of wideband impedance matching, linear phase response, and stable radiation pattern over wideband is essential for the successful transmission and reception of UWB pulses. Many researchers have reported various antenna solutions for UWB communication [6, 7]. Still there is a need of novel antenna solution for pulse transmission. A time-domain analysis is required to compare the performance of these antennas.

Recent research has shown that by using multiple antenna elements at the transmitter and receiver ends known as MIMO system (multiple-input-multiple-output), the channel capacity can be increased without the need of increasing the channel bandwidth or transmitted power level. The limits on data rates have been pushed up by introducing another dimension to mobile communication technology, i.e., spatial diversity, having multiple antenna elements at both receiver and the transmitter ends, the channel capacity increases linearly with the number of antennas in comparison to a conventional single-input-single-output (SISO) system [8]. This system with array of antennas at both transmitter and receiver ends is called the MIMO system. The channel capacity over fading channels having antenna arrays at both the ends was first published by Winters in 1987 [9]. The tremendous potential of MIMO systems was appreciated more than ten years later when they were reinvented by Foschini and Gans [10], and Telatar [11]. Since then, efforts have been put in to overcome challenges of MIMO systems.

Earlier most of the research on MIMO systems focused on the channel properties as well as the signal processing algorithms. The derivations on maximum achievable channel capacity of MIMO systems were based on assuming zero correlation between the signals on different antennas. The new major parameters of focus are the propagation channel and the antenna design keeping in view that signal processing and channel characteristics are equally important. MIMO systems performance improves with lower correlation between the signals on different antennas [12]. Antenna parameters such as (1) array configuration, (2) polarization of antenna, (3) radiation pattern, and (4) mutual coupling between signals impact the correlation.

It is very difficult to accommodate several antennas maintaining low correlation in the small dimensions of the mobile terminal. At the base station, low correlation can be achieved by placing the antennas at appropriate distances ideally half wavelength apart as there is no restriction of space. This is known as spatial diversity. At the mobile terminal, the space is limited so antennas cannot be placed far apart which results in high correlation and, therefore, degrades the MIMO performance. Using diversity techniques such as pattern diversity and polarization diversity, uncorrelated signals can be obtained in a multipath environment by having orthogonal signal patterns. The entire work done in the present book is broadly categorized in two parts, i.e., (1) design and analysis of compact UWB antennas for high data rate communication, and (2) design and analysis of compact MIMO antennas for space crunch applications.

Overall, during the course of the research work, according to the defined problem statement, compact antenna solution has been presented for UWB and MIMO systems to achieve high data rate communication. All above mention issues are presented in Chaps. 1–7 of this book. An introduction of the entire work followed by the motivation behind it and objective of the research is presented in Chap. 1. It also presents the scope of the overall work. Chapter 2 presents a comprehensive review on previous published literature on antenna solution for UWB and MIMO communication systems. The research gap and the problem formulation have been presented. Chapter 3 deals with time-domain analysis of ultra-wideband antennas with notch techniques. In the first section, UWB antenna with single notch has been discussed. In second section, UWB antenna with dual band-notch characteristics has been discussed. In the above two sections, the realized notch is static type notch which cannot be altered in real time. Sometimes, it is required that the notch to be tuned so that a particular target frequency can be rejected. So in third section, UWB antenna with tunable band-notch characteristics has been presented. To mitigate interference with existing narrow band communication systems, UWB antenna with various band-notch techniques has been explored by various researchers in literatures. These techniques result a similar kind of frequency domain notch characteristics, i.e., notch bandwidth and notch strength. However, time-domain performance may get affected differently by different band-notch techniques. A time-domain comparison of band-notch techniques has been discussed in Chap. 4. In Chap. 5, compact MIMO antenna solution utilizing pattern diversity has been discussed for space crunch mobile terminals. The first antenna utilizes a omnidirectional type pattern diversity, whereas the second antenna shows a combined pattern and polarization diversity. In an environment where multiple scattering elements and depolarizing elements are common, use of a circularly polarized antenna gives excellent performance over linearly polarized antenna. Again using a circular polarization diversity MIMO antenna, the channel capacity can be increased. Novel antenna solution for MIMO system utilizing polarization diversity has been designed and analyzed for a compact terminal has been presented in Chap. 6. Finally, Chap. 7 summarizes the contributions made in this book and the scope for the future work is outlined. In summary, this book contributes toward the development of compact antennas for high data rate communications. Wideband antennas for pulse transmission and compact MIMO antennas have been designed, analyzed, and presented in this book.

During the course of the preparation of this book, our colleagues at Indian Institute of Technology Roorkee helped us immensely. We sincerely thank Dr. Jagadish C. Mudiganti, Dr. Ashwini Kumar Arya, Dr. Arjun Kumar, Mr. P. Vamsi Krishna, Mr. V. Paritosh Kumar, Mr. Kumar Goodwill, Mr. Ramesh Patel, Dr. Pravin Prajapati, Mr. Gaurav Singh Baghel, Mr. Sukwinder Singh, Mr. S. Yuvaraj, Mr. Himanshu Maurya, Mr. Ajeet Kumar, Mr. Karan Gumber, Mr. L.D. Malviya, Mr. Ravi Dhakad, Ms. Aditi Purwar, Ms. Diksha Nagpal, and Ms. Priyanka Bansal. Special thanks are due to Indian Institute of Technology Roorkee (IIT–Roorkee) and Ministry Human Resources and Development (MHRD), Government of India, for their generous support and encouragement.

Roorkee, India
March 2017

Jagannath Malik
Amalendu Patnaik
M.V. Kartikeyan

Acknowledgements

We sincerely thank the authorities of IIT–Roorkee for their kind permission to come up with this book. In addition, special thanks are due to all the authors of the original sources for the use of their work in this book.

We sincerely thank the following journals/publications/publishers of conference proceedings for their kind permission and for the use of their works and for reprint permission (publication details are given in the respective references and the corresponding citations are duly referred in the captions):

- IEEE, USA.
- IET, UK.
- Springer+Business Media B.V.
- Wiley Inc.
- EuMA/Cambridge University Press.
- Queen Mary University of London.
- URSI, Belgium.

Contents

Chapter 1
Introduction

1.1 Fundamental Limits in Channel Capacity

There is an ever growing need for increased data rates and enhanced quality of service in performance of the future wireless communication systems. Over the past several years, wireless communication has seen a major growth starting from first generation (1G) analog voice only communication to second generation (2G) digital voice communication and a slower data transmission rate. Now a days, the third generation (3G) mobile communication systems provide digital voice communication with high speed data transfer having applications in Video calls, Mobile TV, Internet access and many other download services. Further, the forthcoming mobile standards like fourth generation (4G), LTE (Long Term Evolution), and WiMAX (Worldwide Interoperability for Microwave Access) aim to provide even higher data rates and longer range with enhanced quality of service to the end users. These are designed for Metropolitan Area Network (MAN) based on IEEE 802.11 standards. It aims to provide high speed wireless internet access over longer distances.

Continuous research efforts have been made by the communication engineers to achieve higher and higher data rates. The upper bound on the maximum achievable channel capacity for the ideal bandwidth limited AWGN (Additive White Gaussian Noise) channel is determined by the Shannon-Nyquist criterion [1, 2]. With an available channel bandwidth, 'W' and signal-to-noise ratio over this bandwidth as 'SNR', the maximum theoretical achievable capacity 'C' is given as:

$$\mathrm{C} = \mathrm{W}\log(1+\mathrm{SNR})\ bits/s \tag{1.1}$$

The increase in transmission rate is possible by raising the RF transmission power and/or by extending the communication bandwidth. The channel capacity increases logarithmically with increase in SNR, i.e. capacity increases with increase in transmit RF power level. Existing wireless systems operate at fixed frequencies and bandwidth with strict regulations of FCC. Higher capacity is attainable only with raising signal power level. A large number of portable devices use battery as its source of power which limits the transmitted power and higher transmitted power leads to the potential interference issue. Consequently, a large frequency bandwidth is the best possible solution to achieve higher data rates. However, the spectrum is pre-

J. Malik et al., *Compact Antennas for High Data Rate Communication*,
Springer Topics in Signal Processing 14, DOI 10.1007/978-3-319-63175-2_1

allocated and the bandwidth can't be used without the prior approval of FCC to accommodate new services. Moreover, increasing bandwidth will cause interference with other narrow-band communication systems. New improved technologies are being developed to introduce new services. These services need to be accommodated with the already existing services without making the RF spectrum overcrowded. It is becoming increasingly difficult to achieve high data communication with existing technologies.

In order to achieve higher data rates with the limited available resources and imposed constraints, wireless communication technology needs to be pushed beyond the physical limits of conventional technology and the wireless propagation channel as well. Fortunately, to address this issue, two promising technologies have been evolved in the past decade, the ultra-wideband (UWB) communication system and the multiple-input-multiple-output (MIMO) communication system. The basic principle of both technologies are distinct and different from conventional communication systems. The approach to achieve higher data rate over Shannon's limit using these technologies is different from one another. The former technology takes the advantage of short time-domain un-modulated pulse transmission, whereas the later takes the advantage of multi-path propagation characteristics of a wireless channel to achieve a higher data rate. These technologies can co-exist with existing narrow-band communication standards without any potential interruption/interference.

1.2 Motivation

Antenna plays an important role in any wireless communication system at the front-end of the transceiver. Overall system performance may degrade if a suitable antenna is not used otherwise. Considering the physical structure and other advantages, printed/microstrip antennas are popular and used for most of the applications ranging from space satellite to modern warfare. Advantages like planar, light weight, conformal to curved structures, and low cost makes it most suitable to be integrated in devices [49–55]. Now-a-days, almost all hand-held mobile devices predominantly uses microstrip antennas.

Research in the field of UWB communications have increased at an enormous rate since the allocation of 3.1 to 10.6 GHz frequency spectrum by FCC, enabling it for commercial use. UWB technology may be the future technology to implement high data rates services. Antenna designers have been showing interest in UWB antennas because of the new challenges and opportunities provided by UWB technology. The need to maintain the characteristics of the antenna as demanded by UWB technology over a huge bandwidth of 7.5 GHz poses a quite significant challenge. Furthermore, small and low profile integrated circuits are desirable for portable electronic devices. This limits the size of the antenna that can be used for UWB systems.

UWB can be used in overlay and underlay communications. It is susceptible to interference from the existing narrow band systems which include Worldwide Interoperability for Microwave Access (3.3–3.7 GHz), IEEE 802.11a WLAN (5.15–

5.35 GHz and 5.725–5.875 GHz) and HIPERLAN/2 (5.450–5.725 GHz), etc. Since the power spectral density of the UWB systems are very low compared to narrow band systems, the interference caused to a narrow band system by a UWB transmitter is negligible. When the narrow band interferer is present close to a UWB receiver, the interference caused is very large. This places very stringent constraints on the filters and the linearity of the UWB front-end. Therefore, a notch at the interfering frequency is required to reduce its effect. Instead of suppressing the interfering frequency with the help of filter, the antenna response itself can be adjusted to reduce the effect. This helps in the reduction of system complexity and cost. Furthermore, because of the dynamic nature of interfering signal, efforts need to be made on designing UWB antennas with tunable notch.

For a while, most of the studies on MIMO technology have focused on signal processing algorithms and channel characteristics. However, the derivations on achievable capacity using MIMO systems assumed ideal antennas with zero correlation between them. While coding and signal processing are key elements to successful implementation of a MIMO system, the propagation channel and antenna design represent major parameters that ultimately impact system performance. As a result, considerable research has been devoted to these two areas. MIMO systems perform best when the correlation between signals on the different antennas is low. Antenna properties such as pattern, polarization, array configuration, and mutual coupling can impact the correlation. It is very challenging task to accommodate multiple antennas with low correlation given the small dimensions of mobile terminals. Conventionally, low correlation can be achieved at the base stations by spacing the antennas an appropriate distance apart, ideally half wavelength distance. This is called spatial diversity. However, at mobile terminals, the space is very limited and the antennas cannot be placed too far apart; this will result in high correlation which would degrade MIMO performance. Pattern and polarization diversities are elegant ways of obtaining uncorrelated signals in a multipath environment by utilizing orthogonal patterns and polarizations respectively. Therefore pattern and polarization diversity based MIMO systems are suitable for implementation on a compact wireless terminal.

Although attempts have been made to find antenna solutions for high-speed communications, a lot more need to be done keeping an eye on the exponential growth in communication industry. This fact prompted us to takeup this task for present research work.

1.3 Research Objectives and Problem Statement

The main objective of the research is to find effective antenna solutions for high-speed communications with special reference to UWB and MIMO system. In order to achieve this objective, the following tasks were attempted and solved successfully.

- Design of compact UWB antennas.
- Implementation of static and tuneable band-notch in UWB antennas.
- Design of compact MIMO antennas with pattern diversity.
- Design of compact MIMO antennas with polarization diversity.

1.4 Organization of Book

The work done in this book aims at giving antenna solutions for high-speed wireless communication systems. The book content can be broadly categorized into two parts, i.e., (a) design and analysis of compact UWB antennas for high-speed communications, and (b) design and analysis of compact MIMO antennas for space constrained scenarios. A brief introduction to limitations on maximum achievable data rate in a AWGN channel, also known as *Shannon's channel capacity* is presented in this chapter. The motivation, objective of the research is presented. It also includes the scope of the overall work. In Chap. 2, a comprehensive review on various aspects of these two communication technologies, namely UWB and MIMO, that enables high-speed communications is presented. Review of antenna solutions for successful implementation of these technologies has also been covered. In Chap. 3, design and analysis of various band-notch UWB antennas is presented. The frequency and time domain analysis of these antennas is also presented. A time-domain comparison of various band-notch techniques is presented in Chap. 4. In Chap. 5, compact MIMO antenna solution utilizing pattern diversity is discussed for space constrained mobile terminals. Antenna solution for MIMO system utilizing polarization diversity is designed and analyzed for a compact terminal in Chap. 6. Finally Chap. 7 summarizes the contributions made in the book and the scope for the future work is outlined.

In summary, the book contributes towards the development of compact antennas for high data rate communications. Wideband antennas for pulse transmission and compact MIMO antennas for hand held mobile devices are designed, analyzed and presented in this book.

Chapter 2
Printed Antennas for High-Speed Communication Systems: Preliminaries and Review

2.1 Ultra-Wideband Communication Systems

2.1.1 Brief History of UWB Technology

UWB technology has its roots that date back to more than a century. UWB technology came to limelight in early 1960s from work on electromagnetic wave propagation study [13]. Historically, UWB systems used the concept of impulse radio where transmission of high data rates is achieved by transmitting a few nano second duration pulses [14]. On 12th December, 1901, spark gap radio transmitter was used by Guglielmo Marconi to transmit Morse code across Atlantic Ocean although he designed it in the late 1890s [4]. Very wide bandwidth pulses were induced by this radio transmitter [15]. The effective recovery of wideband energy and also discrimination from many other wideband signals were not possible at that time. Ultimately wideband communication was abandoned. The communication world turned to narrowband radio transmitter which had the traits of easy regulation and coordination. Efforts were made to lower the interference and to increase the reliability during 1942–1945. This led to filing of several patents on this aspect. Most of the patents were frozen because they were suitable for military applications by the U.S. government [16].

In 1960s, two major substantial breakthroughs in UWB technology took place. An impulse measurement technique to calculate transient behavior of microwave networks was developed by Gerald F. Ross. In 1962, Hewlett-packard developed sampling oscilloscope. This equipment was used to observe, analyze and measure the impulse response of microwave networks and catalyzed methods for generation of sub-nanosecond pulses.

In 1960s, antenna designers developed antennas whose impedance and pattern are almost frequency independent up to certain frequency. Rumsey and Dyson developed logarithmic spiral antennas [17, 18]. The wideband radiating antenna design was carried out by Ross using impulse measurement techniques [19]. This boosted short pulse radar communications. In April, 1973, the short-pulse receiver is the first accepted patent in UWB communications which was filed by Ross [20].

J. Malik et al., *Compact Antennas for High Data Rate Communication*,
Springer Topics in Signal Processing 14, DOI 10.1007/978-3-319-63175-2_2

The unlicensed wideband communication in the bands allocated for Industrial Scientific and Medicine (ISM) was allowed in the mid 1980s. Since then, WLAN and Wi-Fi have grown tremendously. Subsequently the communication industry began to study the merits and implications of wider bandwidth communications.

In 1990, the term 'UWB' was first used in a radar study by Defense Advanced Research Agency [21]. In early days, UWB was considered as 'carrier-free' or 'baseband' signals having extremely fast rise time. Pulse or step signals were used as wideband signal which excites a wide-band antenna. Advancements in semiconductor technology encouraged UWB for commercial applications.

On 14th February, 2002, FCC amended rules that allowed the unlicensed use of 7.5 GHz bandwidth from 3.1 to 10.6 GHz [3]. This allocation led to the substantial growth in research and development of RF circuits and antennas for UWB communications. Industrial investment have increased in this revolutionary technology to provide wide range of services for customers. UWB technology is an emerging technology which is considered as a promising solution to achieve high data rate communication.

2.1.2 General Concepts of UWB Technology

The UWB systems are different from conventional narrow band systems in terms of transmission of information. In conventional narrow band systems, the baseband signal is modulated and transmitted using a carrier frequency. In UWB systems, the transmission and reception of information is done using low duty cycle, short duration (picosecond-nanosecond) pulses. The low power spectral density signals are transmitted over a huge frequency band. Techniques for generation of short pulses that are used in UWB impulse radio communications are discussed in [44–46]. The general representation of narrowband and wideband signals in time and frequency domain is shown in Fig. 2.1.

Even though the power level of the transmitted pulse is very low, concerns regarding the potential interference to other narrow band systems was taken seriously. Therefore, an emission mask which defines the maximum allowable radiated power for UWB devices was specified by FCC. Hand held communication devices, imaging systems and vehicular radar systems are defined as UWB systems by FCC in its first report and order. The spectral mask for these systems (indoor and outdoor) is illustrated in Fig. 2.2. The emission mask for outdoor and indoor applications in the frequency range 3.1–10.6 GHz is −41.3 dBm/MHz. For frequencies higher than 10.6 GHz, the emission mask for outdoor applications is 10 dB lower than the indoor mask [3].

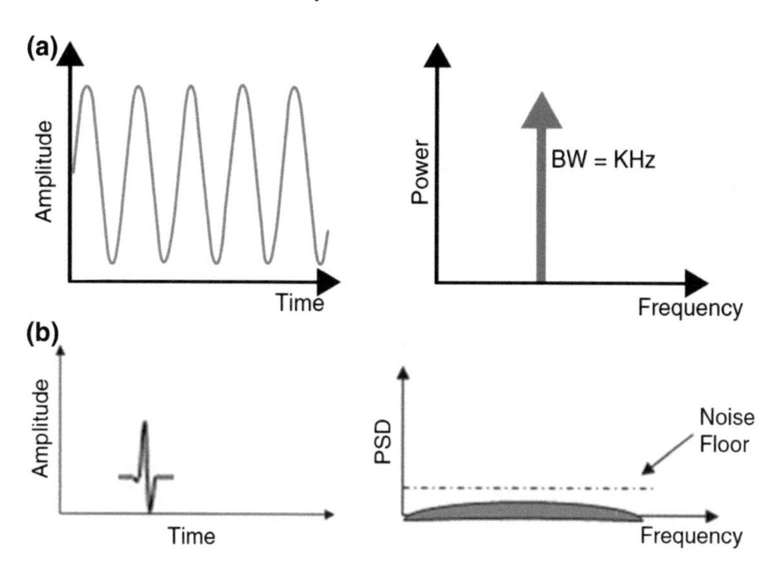

Fig. 2.1 Time-domain and Frequency-domain representation of **a** narrow band signal **b** UWB signal [22] [From: F. Nekoogar and Farid Dowla, Ultra-Wideband Radio Frequency Identification Systems, Springer New York - Heidelberg (ISBN 978-1-4419-9700-5) © Springer Science+Business Media 2011. Reproduced courtesy of the Springer.]

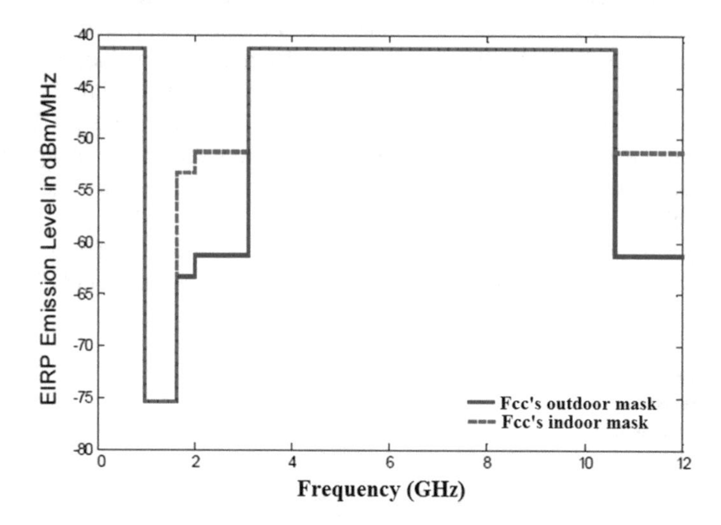

Fig. 2.2 Indoor and outdoor power spectral density emission masks specified by FCC [14] [From: J. Liang, "Antenna study and design for ultra-wideband communication," Ph.D. Thesis, Queen Mary, University of London, United Kingdom, July 2006. Reproduced courtesy of Queen Mary, Uni. of London.]

2.1.3 Advantages of UWB Technology

Ultra-wideband is considered better solution for radar and communication applications because of its unique merits over other technologies. Some of the key advantages of UWB technology are listed below [4, 22].

- Low cost.
- Large channel capacity.
- Low interference to legacy systems.
- Highly immune to multipath interference.
- Less probability of interception and detection.
- Higher accuracy in range measurement and range resolution.

Channel capacity varies linearly with the signal bandwidth at higher frequencies. It is possible to achieve gigabits per second (Gbps) data rate at gigahertz of bandwidth. However, the power limitation on UWB transmission restricts this speed to short ranges up to 10 m [23]. Hence it makes the system more suitable for small range, high data rate wireless applications.

The maximum radiation limit as −41.3 dBm/MHz or 75 nW/MHz makes UWB systems categorized under unintentional radiators. These systems reside well below the noise floor for any conventional narrow-band receiver. Figure 2.3 shows the general idea of coexistence of UWB with narrowband and wideband technologies.

The transmitted signal undergoes multiple reflections from various surfaces and may combine constructively or destructively leading to the distortion of the signal. This mechanism is called as multi-path interference. As the duration of UWB pulse is very short, the chance of collision of reflected pulse with the line of sight (LOS) pulse is very less. Increased immunity to multi-path interference is obtained using UWB signals.

UWB pulses have low average transmission power, hence it appears as noise-like to narrow band communication system. It is very difficult to intercept UWB signal

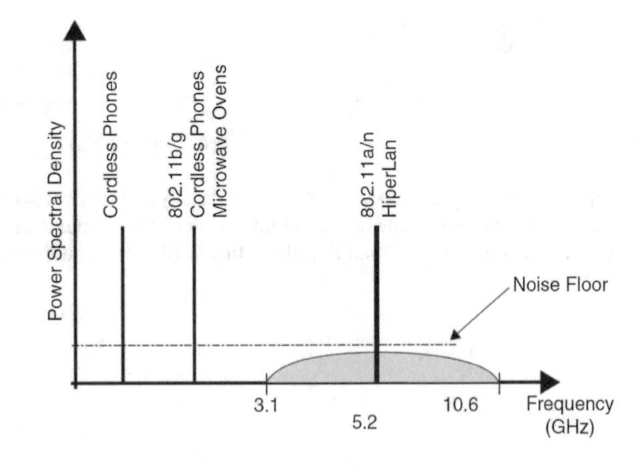

Fig. 2.3 Coexistence of UWB signals with narrowband and wideband signals [22] [From: F. Nekoogar and Farid Dowla, Ultra-Wideband Radio Frequency Identification Systems, Springer New York—Heidelberg (ISBN 978-1-4419-9700-5) © Springer Science+Business Media 2011. Reproduced courtesy of the Springer.]

without a prior knowledge. To secure UWB pulses, unique codes are used to time modulate UWB pulses which are unique to a transmitter/receiver pair [23]. It is impossible to guess a picosecond duration pulse without any prior information. In UWB transmission, modulation of data with a specific carrier frequency is not done, that eliminates the need for RF mixer unlike conventional narrowband technologies [24]. Therefore, fewer RF components are required for carrierless transmission as compared to carrier-based transmission. This decreases the cost and complexity of UWB transceiver.

Higher accuracy in range measurement and range resolution is obtained using a short pulse UWB radar system as compared to a conventional radar system.

2.1.4 Various Applications of UWB Technology

The high data rate characteristic of UWB makes it well suitable to be used in Wireless Personal Area Networks (WPAN) applications. The data is transmitted within a distance of 10 m or less, at rates up to 100–500 Mbps. The wireless connectivity of personal computers to storage devices, printers, scanners, video cameras, using high speed wireless universal serial bus (WUSB) is possible. Download/upload high quality video from mobile devices, audio streaming on MP3 players, wireless high definition digital video receiver, and wireless connectivity of speakers are enabled using UWB in WPANs [25]. The need of cables for interconnection between the mentioned devices is eliminated.

The UWB technology also finds its application in sensor networks. Sensor networks consist of group of sensors distributed spatially in a region for monitoring. The sensor nodes are either static or mobile. The sensor nodes are mobile when equipped on automobiles, firemen, robots, soldiers and emergency response situations [5]. Robustness, multi-functionality, low cost, and low power are the desired properties of sensor networks which can be attained by using UWB technology. An enormous quantity of sensory data can be gathered and disseminated in a timely manner with the help of UWB communication systems. The noticeable drop in installation and maintenance costs is possible without usage of wires.

UWB is an excellent candidate for indoor positioning and tracking applications due to higher data rate at smaller range. The advanced tracking mechanism can be used to track movement of objects in an indoor environment with centimeters of accuracy [15]. They can be used to locate objects or missing people in situations such a burning building, causalities in a remote area and so on [14].

Reflections of UWB pulses from the target causes time shift and amplitude distortion. Scattering affects UWB waveforms significantly as compared to conventional radar systems. Taking advantage of this property, UWB can be used to replace X-ray systems in medical diagnostics, through wall imaging under grounding imaging using ground-penetrating radar for rescue operations.

Resolution of conventional proximity and motion sensors can be improved using UWB-based sensors. The feature of high range accuracy combined with the ability to differentiate targets can be used to build intelligent collision-avoidance systems [5]. Improvement in deployment of airbag and adaptation of breaking/suspension systems with conditions of road are some key applications of vehicular UWB radar systems.

2.1.5 Antenna Requirements for UWB Technology

The design process of UWB antenna is more complex than narrowband antennas as higher number of performance parameters are to be considered in design. The huge bandwidth requirement of UWB antennas makes it different from other antennas. As per FCC regulations, a suitable UWB antenna should achieve at least 500 MHz of absolute bandwidth or at least 20% fractional bandwidth. Generally, the UWB antenna needs to be functional over the entire 3.1–10.6 GHz band.

A consistent performance of UWB antenna over the whole 3.1–10.6 GHz band is required. Ideally, stable radiation pattern, gain and good impedance matching over the entire band is desired. For some applications, in order to make the coexistence possible with other conventional systems, band-rejection characteristic in UWB antenna is required [37–43]. Depending upon the application, an omni-directional antenna or a directional antenna can be used. From application to application, the required radiation property of the antenna also differs. For mobile and hand-held communication systems, an omnidirectional antenna is preferred so that the signal can be received from any direction. The gain of such UWB antenna is of the order of 5 dBi over the entire band. For applications where high gain is required such as radar systems, a directional antenna is used.

The limitation imposed on power spectral density of transmitted pulse is very low in UWB systems. The performance of the system may degrade if losses increase. So, the losses should be as minimum as possible to achieve maximum radiation efficiency. Use of low loss dielectric materials leads to higher radiation efficiency, if a proper impedance matching is maintained at the input port. A greater emphasis is to be laid on size reduction techniques while designing. In order to be compatible with the present day mobile and portable systems, the UWB antenna needs to compact and planar. The antenna should also have the features of low profile and should be compatible with the printed-circuit-board (PCB) technology.

Apart from the above discussed frequency domain performance parameters, a good time-domain characteristics are demanded for UWB antennas. The efficacy of a narrow band antenna is same in its whole bandwidth, while the same cannot be assumed for a UWB antenna because of its huge operational bandwidth. The UWB antenna exhibits significant impact on the transmitted signal. The group delay should be nearly constant over the whole bandwidth, which implies that the phase response should be linear. This minimizes the pulse shape distortion. The ringing duration needs to negligibly small since it lead to inter symbol interference (ISI). The general requirements for a UWB antenna for high-speed communication is given in Table 2.1.

Table 2.1 Antenna parameter considerations for impulse radio communications

Parameters	Requirement
Impedance bandwidth	3.1–10.6 GHz
Radiation pattern	Omni-directional
Phase	Nearly linear
Physical profile	Compact and planar

2.1.6 Characterization of UWB Antennas in Time-Domain

Unlike narrowband communication systems, UWB antennas are used to transmit and receive short pulses (order of pico-seconds) in time domain. Since the antenna radiates over a wide spectral bandwidth, it's necessary to investigate the impulse response of these antennas for successful integration to impulse-radio systems. The parameters mentioned in the previous section are used to characterize UWB antennas in frequency-domain. The time-domain performance of UWB antennas is evaluated based on a variety of parameters. To characterize wideband antennas, some extra parameters are formulated and utilized. These parameters include impulse response, system fidelity factor (SFF) and group delay.

UWB antenna system can be modeled as a two port network as shown in Fig. 2.4 [26]. The antenna system measurement is usually carried out outside anechoic chamber considering a real world situations Fig. 2.5. The antenna system consists of two identical UWB antennas connected to two ports of a vector-network-analyzer (VNA). Keeping in mind of the maximum power output of VNA, the antenna are usually separated by a small distance. To investigation of the system performance, two different spatial orientation between the transmitter and receiver are considered. In one case the antennas remain oriented face-to-face to each other, whereas for second case it is oriented side-by-side to each other. In face-to-face orientation, the two front side of the antennas face each other and in side by side orientation, the front side of the antennas point to the same direction.

The magnitude of system transfer function ($|S_{21}|$) and group delay can be used to analyze the antenna dispersion characteristics. The transmitted pulse will undergo distortion if the phase of S_{21} exhibits nonlinearity at the magnitude drop portion in

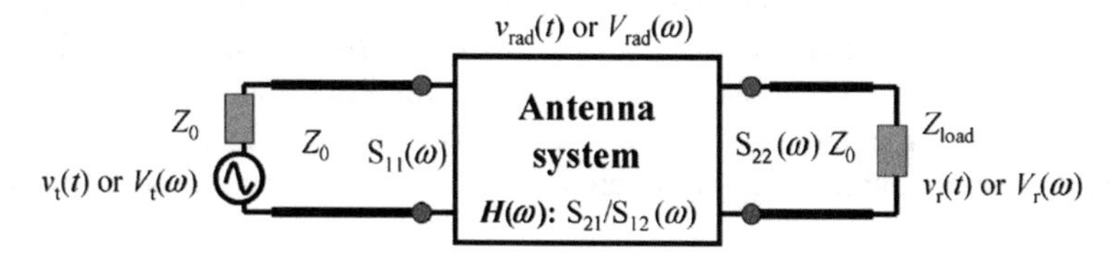

Fig. 2.4 UWB antenna system as a 2-port network [26] [From: Z. N. Chen, X. H. Wu, H. F. Li, N. Yang, and M. Y. Chia, "Considerations for source pulses and antennas in UWB radio systems," *IEEE Transactions on Antennas and Propagation*, vol. 52, no. 7, pp. 1739-1748, July 2004. © IEEE 2004. Reproduced courtesy of IEEE, USA.]

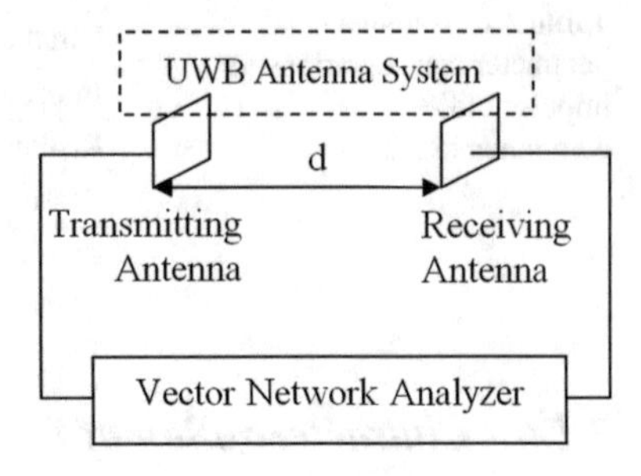

Fig. 2.5 UWB antenna measurement setup using a VNA

the S_{21} plot. But the amount by which the pulse distortion takes place cannot be said with system transfer function alone.

Group Delay parameter is generally encountered when characterizing two port devices. Group delay measures the total phase distortion between the input signal and the output signal. The transfer function $H(\omega)$ of the device is expressed in (2.1) and the group delay is defined as the derivative of the phase response ($\angle H(\omega)$) versus frequency is given in (2.3) as follows:

$$H(\omega) = |H(\omega)|e^{j\angle H(\omega)} \tag{2.1}$$

$$\tau = -\frac{\partial}{\partial\omega}\angle H(\omega) = -\frac{1}{360^0}\frac{\partial}{\partial f}\angle H(f) \tag{2.2}$$

The dispersion produced by the antenna would be less when the transfer function magnitude varies less and the group delay is nearly constant over the desired frequency range. A fairly linear phase of S_{21} and small the group delay variation (<1 ns) over the entire UWB band ensures pulse transmission with minimal distortion.

The impulse reponse of the antenna system can be obtained using the method of Hermitian processing as described in [27]. First, the pass-band signal can be obtained with zero padding from the lowest frequency measured in VNA down to DC. The complex conjugate of the signal can be obtained and when reflected to the negative frequencies gives a double sided spectral response. This is done to obey the property of time domain real signals that the Fast Fourier Transform (FFT) of real signals is conjugate symmetric in nature. Hence the Inverse Fast Fourier Transform (IFFT) of the complex conjugate signal gives real valued signal. The resultant double-sided spectrum after reflection corresponds to a real signal spectrum. Then IFFT of the double sided spectrum is performed to obtain the impulse response in time domain. The algorithm for obtaining impulse response is shown in Fig. 2.6. The received signal can be obtained by simply doing convolution of impulse response with the transmitted signal.

The transmitted signal is subjected to dispersion by the antenna as well as the channel. The fidelity factor quantifies the amount of resemblance between the input driving voltage and the radiated waveform of a transmitting antenna [28]. The correlation between the normalized input signal and the normalized radiated field gives

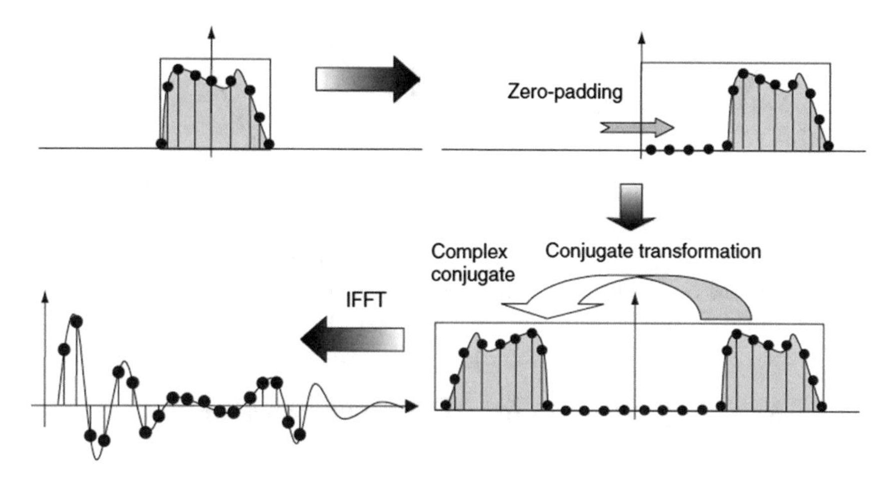

Fig. 2.6 Zero padding, conjugate reflection and resulting impulse response [27] [From: I. Oppermann, M. Hamalainen, and J. Iinatti, UWB Theory and Applications, John Wiley & Sons Ltd., 2004. Reproduced courtesy of John Wiley, USA.]

fidelity factor [29]. In practical scenario, it is difficult to measure radiated field. Alternately SFF is defined as correlation between the transmitted signal and the received signal. From the S_{21} parameter of the antenna system (transmitting antenna, channel, receiving antenna), SFF quantifying the extent of resemblance between the transmitted signal and the received signal. Fidelity factor considers the distortion yielded by transmit antenna only while the system fidelity factor considers the distortion yielded by two antennas (transmitting antenna and receiving antenna) [30]. The correlation between the normalized transmitted and the normalized received pulse gives SFF as:

$$SFF = max_{\tau} \frac{\int_{-\infty}^{\infty} T_s(t).R_s(t+\tau)dt}{\left[\int_{-\infty}^{\infty} |T_s(t)|^2\, dt\right]^{1/2} \left[\int_{-\infty}^{\infty} |R_s(t)|^2\, dt\right]^{1/2}} \tag{2.3}$$

Normalization is performed so that only the shape of the signals is taken into account without considering magnitudes, since the received signal magnitude is always less than transmitted signal [79]. The value of SFF lies between 0 and 1. The received signal is entirely different from transmitted signal when the SFF value is 0, and is identical when the SFF value is 1. The signal is unrecognizable when the distortion is greater than 50% or SFF is <0.5.

Ringing is defined as undesired oscillation of in signal. Ringing effect in antennas arise due to the stored energy or multiple reflections between the input port and the antenna [31]. The time taken by the envelope of a signal h(t) to reduce from peak value to a certain lower value (5% of the peak) is called the ringing duration.

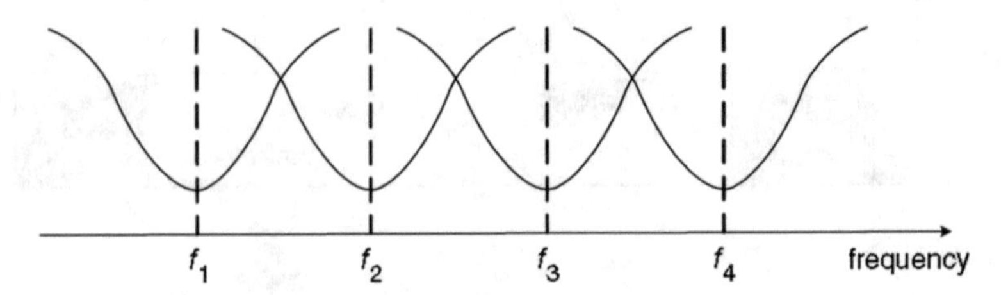

Fig. 2.7 Overlapping of multiple resonance modes [14] [From: J. Liang, "Antenna study and design for ultra-wideband communication," Ph.D. Thesis, Queen Mary University of London, United Kingdom, July 2006. Reproduced courtesy of Queen Mary Uni. of London.]

2.1.7 Revisiting Antennas for UWB Communications

The huge impedance bandwidth of planar monopole antennas is achieved by overlapping of closely spaced resonance modes. General illustration of this phenomenon is shown in Fig. 2.7.

An extensive study on the performance characterization of a CPW fed circular disc monopole antenna has been conducted in [127]. The critical parameters that define the performance of the antenna are the feed gap, width of ground plane and radius of monopole disc. Since the first two parameters depend on the ground plane, it plays a significant role that affect the operational bandwidth and input impedance matching. It acts as an impedance matching circuit. The UWB characteristic is achieved by the overlapping of multiple resonances and harmonics.

A CPW fed planar monopole antenna which is in the shape of L-I is presented in [128]. The wideband operation is achieved using the concept of resonance overlapping. The I-shaped open stub and an L-shaped monopole are connected to the center metal of the CPW feed line. The L-shaped monopole antenna excites resonant frequencies at 4.1 and 9.2 GHz while the I-stub excites resonant frequency at 5.4 GHz. The three resonant frequencies overlap to attaining a 3–11 GHz impedance bandwidth. In [84], a UWB antenna with huge bandwidth from 2–20 GHz is designed using three pairs of multi-resonant split ring loops. The resonances of these split ring loops overlap leading to the huge bandwidth.

To improve impedance matching between the feeding structure and radiator, an impedance matching network can be employed. It can be done in two ways. Introducing an on-patch matching network by cutting out the lower part of the patch near the feed in various shapes. An impedance matching network can be introduced in the feed line without modifying the radiator. The on-patch matching network transfers the high impedance at the radiating edge of the patch to the 50 Ω characteristic impedance of the feed. Suitable step or linear tapering on the lower part of the patch near the feeding structure leads to the gradual variation of impedance instead of sudden variation. The impedance matching is considerably increased at higher frequencies because of the enhancement in traveling wave modes. Stepped trimming at the lower side of the patch is done forming two steps [129], and four stepped for wide bandwidth [130]. In [72], lower side of the patch is trimmed linearly to achieve wide band operation. Patch in the shape of a bell is designed for impedance matching over a wide bandwidth [131].

Stepped impedance transformer acting as an impedance matching network in the feeding structure to achieve wideband impedance bandwidth is presented in [132, 133]. In [134], a linearly tapered impedance transformer is used in a coplanar-fed monopole antenna for impedance matching.

The antennas impedance bandwidth depends on the ground plane because of the coupling effect between the ground and lower side of the radiating patch. The ground plane is also a part of impedance matching network and it also contributes to radiation. The ground plane edge lying below the patch can be step tapered, linearly tapered or exponentially tapering leading to the reduction of capacitance between the bottom of the patch and the ground. This improves the impedance bandwidth of the antenna.

In [135], the outer part of the ground plane is trimmed that resulted in an inverted stair-style ground. In [136], the ground plane is linearly tapered and the matching at higher frequencies is considerably increased with no significant effect at the lower frequencies. After linearly tapering the outer part of the ground plane near the lower side of the radiating patch, antennas impedance bandwidth is increased by more than two times. Exponentially tapered ground plane is discussed in [138]. Because of tapered ground, multi resonance characteristics are obtained at around 5, 10 and 16 GHz. In [137], a combination of above mentioned techniques have been used and a compact CPW fed UWB antenna having an impedance bandwidth 8.4 GHz from 3–11.4 GHz was designed.

A trident shape feeding strip to achieve linear surface current on patch is presented in [139]. This results a wider bandwidth compared to a simple feeding line. Impedance bandwidth have been increased by more than two folds. When the planar monopole antenna is fed by multiple lines, vertical current distribution increases and the distribution of current in horizontal direction is decreased, leading to the improvement in achieved impedance bandwidth [136].

A full disc UWB monopole antenna was designed in [140]. Then the antenna has been cut across the symmetrical plane and the feed line has been tapered to achieve huge bandwidth from 2.42–13.62 GHz exhibiting stable gain and omni-directional radiation patterns. 50% reduction in size was achieved due to symmetric cut.

Dielectric resonator antennas (DRA) are also investigated for wideband applications [56–58, 61]. These antennas are compact in nature and use a high dielectric material as the radiating element with consistent radiation pattern, good radiation efficiency and gain. Patch antennas with fractal geometries are also investigated for multi-band and wideband applications [68–70].

An UWB antenna can be made to reject or suppress certain bands by etching slots on the patch and/or on the ground plane. Destructive interference takes place when the total slot length is adjusted to be approximately $\lambda_g/2$ at the notch frequency desired, hence the antenna is non-radiating at that frequency. The currents on either side of the gap formed are out of phase and hence destructive interference takes place [141]. The effective length of the slot can be calculated by summing the lengths of all the segments that form the slot. Therefore, the notch frequency is dependent on the length of the slot. The gap of the slot can be altered to change the rejected bandwidth. The location of slot on the patch, the length and gap of the slot are the critical parameters. Multiple slots on the patch produce multiple notch bands. An E-shaped slot in [142], C-shaped slot in [143] were used for producing band-notch. A modified π-shaped slot and a half arc shaped slot have been etched on the same patch for producing two notch bands [144].

A slot in the shape of 'L' has been etched in the ground plane of a CPW fed antenna to create notch at 5.5 GHz [145]. The length od the slot is approximately $\lambda_g/4$ at the notch frequency. The stop-band bandwidth changes by varying the location and width of the slot. At the notch frequency, it was observed that the surface current is concentrated only around the slot. In [146], quarter wavelength slots have been etched on the patch to obtain band-notched characteristics at 3.3–3.7 GHz, 5.15–5.825 GHz and 7.25–7.75 GHz. In the equivalent transmission model, each slot can be considered as a shorted stub connected in series with the primary transmission line. At the notch frequency, as its length becomes quarter wavelength, infinite input impedance appears in series between the input transmission line and the antenna. So the radiating patch and the feed line becomes impedance mismatched [147].

In [73], the elliptical radiator has been connected with two resonating inverted L- shaped stubs to produce band rejection characteristic. The L- shaped stub length is approximately $\lambda_g/4$ at notch frequency. An open stub can be connected to the primary transmission line as equivalent for L-shaped stub. At the notch frequency, the stub acts as short on transmission line with no power reaching the antenna, hence no radiation from the antenna.

The quarter wavelength stub can be of any shape. In [74], a meander shaped quarter wavelength stub is used to produce band-notch characteristic. In [72], a parasitic loop is placed beneath radiating patch. The parasitic loop length is $\lambda_g/2$ at notch frequency. The currents on the radiating patch and parasitic loop are opposite to each other at the notch frequency [43], hence band-notch characteristic is obtained. In [148], SRR array is placed on the back side of the UWB antenna to achieve band-notch characteristics. It is observed that the SRR array position affects the higher frequency resonance characteristics. The SRR array acts as cascaded parallel LC circuit. The input signal sees high input impedance at SRRs resonant frequency. So the radiating patch and the feed-line are impedance mismatched [148].

In [75], an UWB antenna with triple band-notch characteristics is designed. A radial slot in ground plane, a pentagonal slot etched on the patch and a pair of stubs connected to the feed-line below the radiating patch are used to obtain band-notch performance. The stub length is quarter wavelength at notch frequency and in the equivalent circuit model it acts as a short circuit. Hence no power reaches the antenna at that frequency.

Varactors have been used to achieve the tunability in notch frequency. In [94], a quarter wavelength open stub resonator is placed under the transmission line and is connected through a 'via' hole to the transmission line. Stubs are loaded with varactors. The capacitance of Varactor changes with the applied bias voltage, hence the electrical length of the stub changes. Consequently the notch frequency is changed. In [149], the slot etched on the UWB antenna, and loaded with a varactor is used to achieve the tunability of the notch frequency in the band 5–6 GHz. The bias circuit is designed using distributed elements. In [150], open loop resonators are placed on either side of the microstrip feed. A varactor is placed in the center of the resonator to tune effective electrical length of the loop to achieve tunability in the notch frequency.

2.2 Multiple-Input-Multiple-Output Communication Systems

2.2.1 Brief Description of MIMO Communication Systems

In a conventional radio system, there is single antenna used at transmitter and at receiver. This system is called a single-input-single-output (SISO) System. The capacity of such systems is limited by Shannon-Nyquist criterion. In order to increase channel capacity of the SISO systems to meet increasing demands for high-speed communications, the bandwidth and transmission power have to be increased significantly. However it is not a viable solution as discussed earlier. Recent developments have shown that using MIMO systems could substantially increase the capacity in wireless communication without increasing the transmission power and bandwidth [9, 10]. Rayleigh fading caused by multipath has been a source of problem for conventional wireless systems. However, to increase the channel capacity, MIMO systems exploit multipath instead of mitigating it. MIMO wireless systems have demonstrated the potential of increased capacity in rich multipath environments. Such systems operate by exploiting the spatial properties of the multipath channel, thereby offering a new dimension which can be used to enhance performance. A typical MIMO system is shown in Fig. 2.8 with multi channel concept. The performance improvements by using MIMO systems are due to array gain, diversity gain, spatial multiplexing gain, and interference reduction [8]. A brief description of these is given below:

1. *Array Gain*: Array gain is the enhancement of signal quality at the receiver through signal processing. The SNR at the receiver can be improved by combining coherently the signals received at each antenna [8].
2. *Diversity Gain*: Diversity is a powerful technique to mitigate fading in wireless links. Diversity techniques rely on transmitting the signal over multiple independently fading paths (in time/frequency/space) [8]. The received signals can be combined together in some optimal way to yield a signal with better SNR.
3. *Spatial Multiplexing Gain*: MIMO channels offer an increase in capacity for no additional power or bandwidth expenditure. This gain, referred to as spatial multiplexing gain, is realized by transmitting independent data signals from the individual antennas. Under favorable channel conditions, such as rich scattering, the receiver can separate the different streams, yielding an increase in the capacity [8].

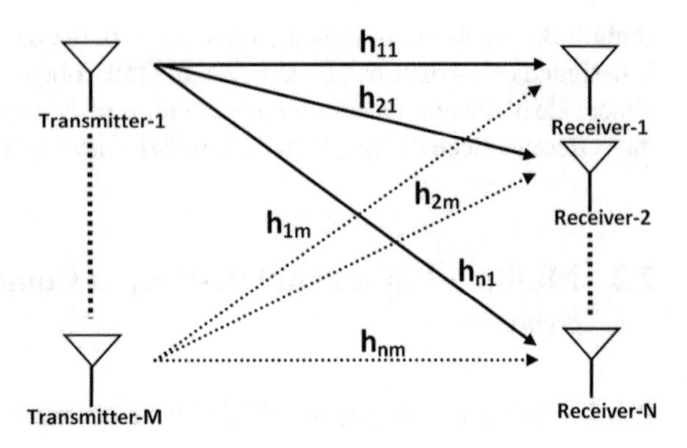

Fig. 2.8 A MIMO system comprising M-transmit antennas and N-receive antennas

4. *Interference Reduction*: When multiple antennas are used, the differentiation between the spatial signatures of the desired signal and interference signals can be exploited to reduce interference. Interference reduction requires knowledge of the desired signals channel. Exact knowledge of the interferers channel may not be necessary [8].

2.2.2 Channel Capacity in MIMO Systems

The channel capacity of a MIMO system with 'N' transmitting and 'N' receiving antenna is discussed here. Figure 2.8 shows a MIMO system with 'M'—transmitting antennas and 'N'—receiving antennas. For generalization, consider M=N for the present case.

For a narrowband channel, the complex transmission coefficients between element $k \in [1, \ldots, N_T]$ at the transmitter and element $j \in [1, \ldots, N_R]$ at the receiver at time 't' is represented by $h_{jk}(t)$. A matrix containing all channel coefficients (channel coefficient matrix, H(t)) can be shown as:

$$H(t) = \begin{pmatrix} h_{1,1}(t) & h_{1,2}(t) & \ldots & h_{1,N_T}(t) \\ h_{2,1}(t) & h_{2,2}(t) & \ldots & h_{2,N_T}(t) \\ \ldots & \ldots & \ldots & \ldots \\ h_{N_R,1}(t) & h_{N_R,2}(t) & \ldots & h_{N_R,N_T}(t) \end{pmatrix} \tag{2.4}$$

Hence, a system transmitting the signal vector $x(t) = [x_1(t), x_2(t), \ldots, x_{N_T}(t)]^T$, where $x_k(t)$ is the signal transmitted from the kth antenna would result in the signal vector $y(t) = [y_1(t), y_2(t), \ldots, y_{N_R}(t)]^T$ being received, where $y_j(t)$ is the signal received by the jth antenna, and

$$y(t) = H(t)x(t) + n(t) \tag{2.5}$$

where n(t) is the noise vector. MIMO systems can achieve orthogonal sub channels between the transmitters and receivers through a rich scattering environment and consequently increase the offered capacity. Mathematically, the number of independent sub channels can be estimated by using the singular value decomposition (SVD) of the channel coefficient matrix H as described in [12] as:

$$H = USV^H \tag{2.6}$$

where $(.)^H$ denotes Hermitian operator. Let the unencoded $N_T \times 1$ transmit vector in the waveform domain be denoted as x'. We encode the transmit vector as $x = Vx'$. Since each element of x' multiplies the corresponding column of V, this operation suggests that each column of V represents array weights for each signal stream. The receiver performs the operation $y' = U^H y$. Indicating that each row of U^H represents the receive array weights for each stream. Because U and V are unitary,

$$y' = U^H y = Sx' + \eta' \tag{2.7}$$

where $\eta' = U^H \eta$. Since the matrix S of singular values is diagonal, (2.7) indicates that y' is a scaled version of the transmit vector corrupted by additive noise. Therefore, this processing has created independent (spatially orthogonal) parallel communication channels in the multipath environment. The number of independent parallel communication channels is equal to rank of H matrix. Therefore in rank deficient channels, the number Q of independent streams should correspond to the number of singular values that are above the noise floor. In this case, only the first Q columns of U, S and V should be used in the processing and analysis.

To see how parallel spatial channels can increase capacity, consider the simple case of Q uncoupled transmission lines. If only one transmission line is used to send data, the Shannon channel capacity will be

$$C = log_2(1 + \rho) \tag{2.8}$$

where ρ is the receiver SNR. If the transmit power is divided equally among the lines, the capacity becomes [12]:

$$C_Q = \sum_{q=1}^{Q} log_2(1 + \rho/Q) = Qlog_2(1 + \rho/Q) \tag{2.9}$$

where equal receiver noise is assumed. For a transmit vector whose elements are complex Gaussian-distributed random variables, the expression for channel capacity is [12]:

$$C = MAX_{R_x:Tr(R_x)\leq P_t} log_2 \left| I + \frac{HR_xH^H}{N_T\sigma^2} \right| \tag{2.10}$$

where $R_x = E(xx^H)$ is the transmit covariance matrix. The diagonal elements of R_x represent transmit power from each antenna and hence the constraint $Tr(R_x) \leq P_t$, where P_t is the total transmit power. Determining the capacity involves identifying the covariance matrix R_x that maximizes (2.10). However, when the transmitter does not know H, it can divide power equally among the transmit antennas to form N_T independent streams or $R_x = (P_T/N_T)I$. Thus the uninformed transmit capacity is given by [12],

$$C = MAX_{R_x:Tr(R_x)\leq P_t} log_2 \left| I + \frac{HR_xH^H}{N_T\sigma^2} \right| \tag{2.11}$$

$$C_{UT} = log_2 \left| I + \frac{P_T HH^H}{N_T\sigma^2} \right| \tag{2.12}$$

which may be decomposed as [8]:

$$C_{UT} = \sum_{i=1}^{r} log_2 \left(1 + \frac{P_T\lambda i}{N_T\sigma^2} \right) \tag{2.13}$$

where r is the rank of H and λ_i (i = 1 to r) are the positive eigenvalues of HH^H. It can be seen from (2.13) that multiple scalar spatial data pipes (also known as spatial modes) open up between transmitter and receiver resulting in significant capacity gains over the SISO case. For example, C_{UT} increases by r bit/s/Hz for every 3-dB increase in transmit power (for high transmit power), as opposed to 1 bit/s/Hz in conventional SISO channels. It has been established that at high SNR, the ergodic capacity of fading MIMO channels is [8]:

$$C = min(N_R, N_T) log_2 \rho + O(1) \tag{2.14}$$

This clearly shows that the capacity increases linearly with the minimum of the number among transmit and receive antennas.

2.2.3 Signalling Schemes in MIMO Systems

There are basically two signalling schemes for MIMO systems.

1. *Spatial Multiplexing*:
   In spatial multiplexing (SM) independent data streams are transmitted simultaneously in parallel channels from each element in an array of antennas. The basic principle of SM is illustrated in Fig. 2.9, wherein a system with two elements at the transmitter and two elements at the receiver is considered [32]. Firstly, the bit stream of data to be transmitted is de-multiplexed into two sub streams, then modulated and transmitted simultaneously from each transmit antenna as

shown in Fig. 2.9. Under favorable channel conditions such as rich scattering, these signals arriving at both the receiving antennas are well separated. The the receiver has knowledge of the channel, hence it can differentiate between the co-channel signals and extract both signals. After demodulating the received signals, the original sub streams can be combined to yield the original bit stream of data. Therefore, spatial multiplexing leads to increase in the channel capacity with the number of transmitting-receiving antenna pairs. This concept can be extended to more general MIMO channels.

The maximum number of parallel channels that can be achieved in an ideal MIMO system is min(N_T, N_R). Shannons capacity formula for SISO system indicates that in the high SNR regime, a 3 dB increase in SNR will approximately increase capacity by 1 bit/s/Hz. However, in the MIMO systems, the capacity in the high SNR regime will increase by min(NT ,NR) bits/s/Hz with every increment of 3 dB in SNR. However, SM does not work well in low SNR environments as it is more difficult for the receiver to identify the uncorrelated signal paths [32].

2. *Space-Time Coding*:
An alternative method of exploiting MIMO channels is known as space-time coding. It aims to improve the systems performance by using multiple antennas for diversity gain rather than for the spatial-multiplexing gain. It increases network throughput by selecting quality signal paths such that higher data rates can be achieved [32].

In space-time coded transmitter, a single data stream is encoded across both time and space to produce the symbol streams for each transmitter as shown in Fig. 2.10. Appropriate decoding at the receiver allows a diversity gain to be achieved. This

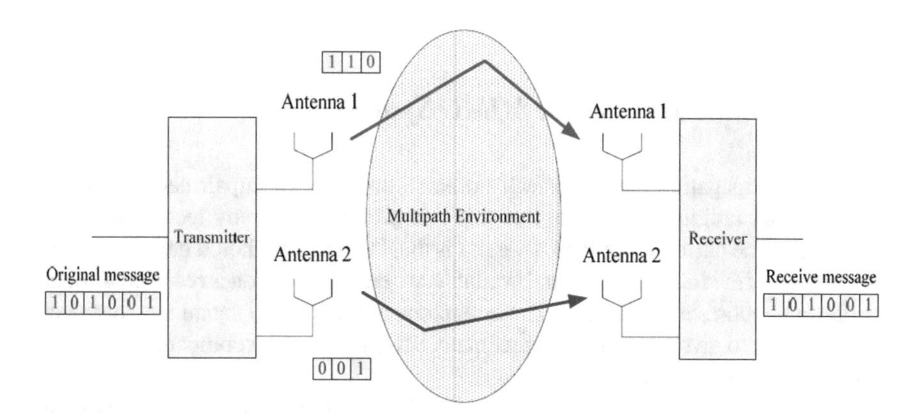

Fig. 2.9 A 2 × 2 MIMO system with a spatial multiplexing scheme [32] [From: C. Chiau, "Study of the diversity antenna array for the MIMO wireless communication systems," Ph.D. Thesis, Department of Electronic Engineering, Queen Mary, University of London, United Kingdom, April 2006. Reproduced courtesy of Queen Mary, Uni. of London.]

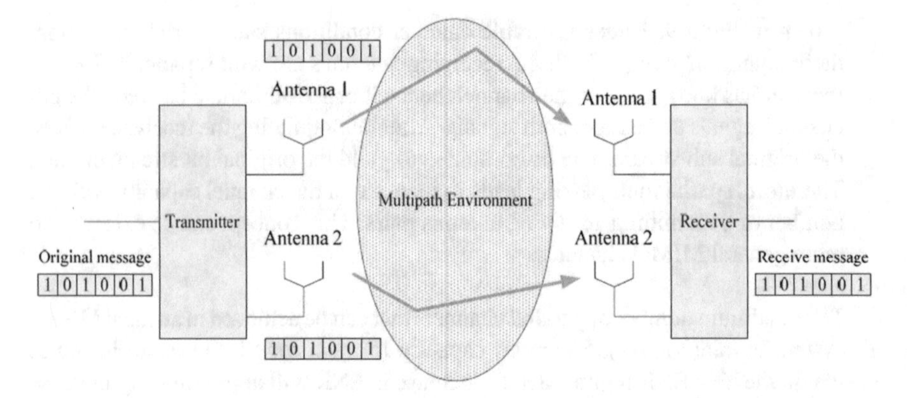

Fig. 2.10 A 2 × 2 MIMO system with a space-time coding scheme [32] [From: C. Chiau, "Study of the diversity antenna array for the MIMO wireless communication systems," Ph.D. Thesis, Department of Electronic Engineering, Queen Mary, University of London, United Kingdom, April 2006. Reproduced courtesy of Queen Mary, Uni. of London.]

method is particularly attractive as it does not require channel knowledge at the transmitter [32]. The resulting diversity gain improves the reliability in a fading wireless link, hence it improves the quality of the transmission. It should be noted that space-time coding scheme does not increase the capacity linearly with the number of transmit/receive elements used. However, it maximizes the wireless range and coverage by improving the quality of the transmission. In order to improve both range and capacity, a MIMO system requires to support both the spatial multiplexing and space-time coding schemes. Examples of diversity coding techniques are the Alamouti coding scheme and delay diversity.

2.2.4 *Antenna Diversity in MIMO Systems*

Multipath propagation had historically been regarded as an impairment because it causes signal fading. In order to mitigate this problem, diversity techniques were developed. The basic principle of diversity is that the receiver should have more than one version of the transmitted signal available, where each version is received through a different uncorrelated channel. These versions of transmitted signal are combined appropriately to give signal with higher mean SNR at the output compared to a single branch resulting in a diversity gain.

Figure 2.11 shows that a dual-element diversity antenna at a receiver that can receive two different versions of transmitted signals and combine them using a combining circuit. There are three kinds of diversity combining techniques

1. *Selection Diversity*: Chooses the path with the highest SNR, and performs detection based on the signal from the selected path.

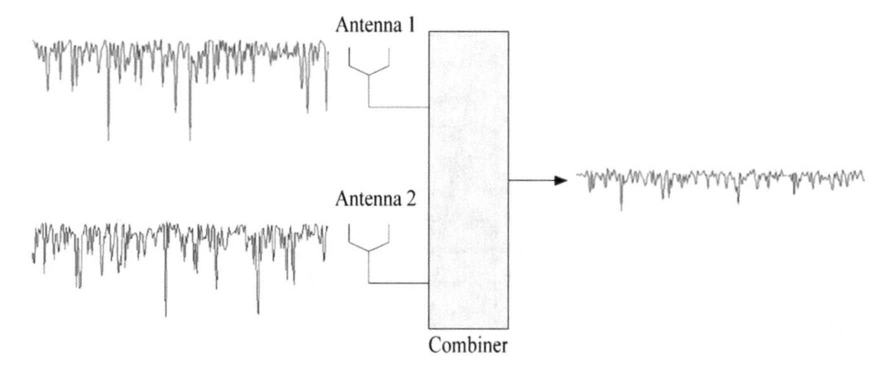

Fig. 2.11 Two signals are combined in a basic diversity receiver [32] [From: C. Chiau, "Study of the diversity antenna array for the MIMO wireless communication systems," Ph.D. Thesis, Department of Electronic Engineering, Queen Mary, University of London, United Kingdom, April 2006. Reproduced courtesy of Queen Mary, Uni. of London.]

2. *Maximal Ratio Combining* (*MRC*): Makes decisions based on an optimal linear combination of the signals.
3. *Equal Gain Combining* (*EGC*): Simply adds the signals after they have been co-phased [33].

MIMO systems exploit the channel spatial degrees of freedom to increase communication performance. In traditional antenna diversity, these resources are used to transmit and/or receive duplicate copies of a single information stream to increase the reliability of detection. In contrast, spatial multiplexing indicates sending distinct information streams over the channels to increase throughput and spectral efficiency.

The combined diversity and spatial multiplexing accomplished with a MIMO system will depend on the required throughput and quality-of-service for an application. This relationship implies that the traditional mechanisms used in diversity systems for reducing antenna correlation generally work to improve MIMO performance as well. It is emphasized, however, that low correlation is a necessary but not sufficient condition for good MIMO performance, since the propagation environment must also possess the appropriate characteristics [34].

It must be noted that if the mean powers in different signal branches are not in same level, then signal in a weaker antenna may not be useful although it is less faded. Therefore to design antenna with a good diversity, the mean power levels in all signal branches should be similar and the correlation between between antennas must be minimum. There are five categories of diversities, i.e. (a) frequency diversity, (b) time diversity, (c) spatial diversity, (d) pattern diversity and (e) polarization diversity. Amongst these five diversities, only the spatial, pattern and polarization diversity techniques are categorized as antenna diversity [32]. These antenna diversity techniques can be applied to obtain uncorrelated signals in MIMO system. The following subsections describe these three types of antennas diversities:

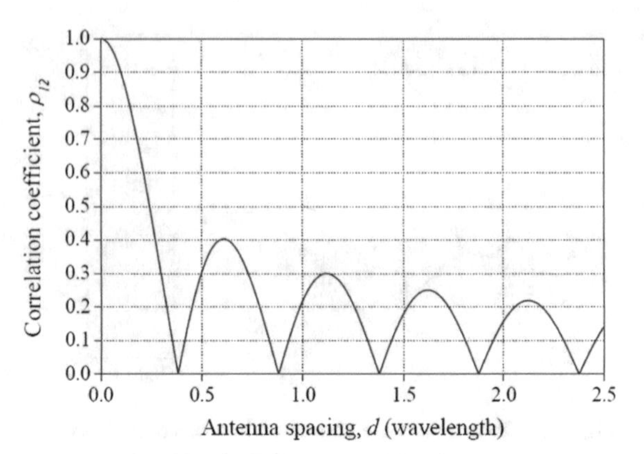

Fig. 2.12 The effect of inter element spacing on correlation coefficient [32] [From: C. Chiau, "Study of the diversity antenna array for the MIMO wireless communication systems," Ph.D. Thesis, Department of Electronic Engineering, Queen Mary, University of London, United Kingdom, April 2006. Reproduced courtesy of Queen Mary, Uni. of London.]

1. *Spatial Diversity*:
   In spatial diversity more than one antenna are employed which are sufficiently separated from each other so that the relative phases of the multipath contributions are significantly different at the two antennas. When large phase differences are present, they give rise to a low correlation between the signals at the antennas. By assuming a two dimensional scenario wherein angular density function is taken to be uniform in azimuth of the mobile environment and no angular density function in elevation, the correlation coefficient for a distance separation 'd' can be obtained from the *zero*th order Bessel function, $J_0(x)$ [32].
   The first null of is $J_0(\beta d)$ is at d = 0.4λ, as shown in Fig. 2.12, where β is the phase constant. As shown graphically in Fig. 2.12, the correlation coefficient has a minima at d = 0.4λ and increases thereafter. However, in suburban areas the measurements show that the first null appears at about d = 0.8λ. This may be due to a lack of uniform angular distribution of wave arrival [32]. Generally, spacing, d of 0.5λ is practically used to obtain two uncorrelated signals at mobile terminals.
2. *Pattern Diversity*:
   Pattern diversity occurs in many instances at the mobile terminals because the antennas will pick up signals coming from different angles. Since the fading signals coming from different directions in a multipath environment are independent, use of pattern diversity can lead to uncorrelated channels. At the mobile terminal, two omni-directional antennas interacting with each other whilst closely spaced can also obtain pattern diversity. Basically, the antennas act as parasitic elements to each other and their patterns change to allow signals to be picked up at different angles. Recent studies conducted on pattern diversity in the MIMO systems have shown that with appropriate dissimilarity in the antenna pattern, the system can achieve large channel capacity [34]. One suggested approach for realizing such a situation involves the use of multimode antennas where the patterns for different modes exhibit high orthogonality (low correlation) [35]. Finding other antenna topologies that offer this orthogonality in a compact form remains an area of active research [59, 60].

3. *Polarization Diversity*:
   Recent work has suggested that in a rich multipath environment, sensing the three Cartesian vector components of the electric and magnetic fields can provide six uncorrelated signals at the receiver. With the use of polarization diversity the size of the antenna structure can be reduced significantly. Polarization diversity can potentially lead to low correlation on at least two branches even when the channel is characterized by little or no multipath scattering. To understand the fundamentals of polarization diversity in MIMO systems, consider two dipoles that are rotated by angle α against one another, are used at the transmitter and at the receiver. Both transmit antennas radiate the same power. For increasing α capacity increases, since the correlation among the channel coefficients of H (the channel matrix) decreases. For $\alpha = 90^0$, the capacity is maximum as two different channels for two different polarizations exist. Those channels are coupled by cross polarization coupling and are not necessarily orthogonal sub channels. The correlation among the two signals is not only influenced by polarization diversity, but also by pattern diversity. Polarization and pattern diversity are always combined and can not be exploited separately [36].

2.2.5 Revisiting Antennas for MIMO Communications

Multiple-input multiple-output (MIMO) systems have emerged as a very interesting strategy to increase the capacity of wireless systems in rich scattering environments [35, 151]. Traditionally, a MIMO system employs several transmit and receive antennas at each end of the radio link, and in order to achieve a high capacity, different signal paths between them should be uncorrelated. The MIMO gains are mainly due to the fact that a rich scattering environment provides independent transmission paths from each transmit antenna to each receive antenna. Antenna diversity is a well-known technique to enhance the performance of wireless communication systems by reducing the multipath fading and co-channel interference [152, 153].

Antenna diversity can be realized in several ways [62–64]. Depending on the environment, space and the expected interference, designers can employ one or more of these methods such as spatial, polarization and pattern diversities to achieve diversity gain. Spatial diversity is employed using spatially separated antenna elements to achieve diversity gain [154–156]. In case of small portable devices, the available space is very limited and it is hard to implement more than one antenna element in such a small space.

However, degradation would result upon placing the antenna elements in close proximity, due to near-field effects, diffraction from finite-ground planes, and strong inductive and capacitive coupling between the elements [120, 152, 157].

The isolation between two antennas or two ports of a single antenna is a critical parameter in diversity antennas and MIMO systems. If the two ports/antennas are strongly coupled then most of the transmitted signal of one port/antenna will not be radiated, but rather received by the second port/antenna which will lead to a reduction of the radiation efficiency due to the dissipation of power in the coupled port/antenna. Therefore, for the two ports/antennas to radiate efficiently they should be sufficiently isolated and the mutual coupling should be small.

In addition, for application in portable devices very small form factors are an important requirement and therefore good isolation between antennas with closely packed antenna elements is necessary. For this reason, polarization and pattern diversity techniques are more suited in the portable devices wherein we can place two antenna elements very close to each other or a single element with more than one feed to achieve diversity gain.

In literature, different techniques have been introduced to increase isolation and reduce the mutual coupling between the two ports/antennas. For example, in [122, 158, 159], modifications on the ground plane have been reported with a view to increase the isolation. An alternate way of reducing the electromagnetic coupling between antenna elements sharing a common ground plane or chassis is to introduce resonant defects or slits in the ground plane. By proper choice of dimensions, the slits resonate and can trap some of the energy between the radiating elements [160, 161]. These resonant slots are designed to be either $\lambda_g/4$ or $\lambda_g/2$ in electrical length depending on open ended or closed type slots. These basically filters/blocks the co-polarized coupling current.

Another strategy to mitigate the coupling between radiating antennas is to use periodic and electromagnetic band-gap (EBG) structures. EBG structures were used extensively in mutual coupling reduction in planar and low-profile antennas [162–166]. These are basically high impedance surfaces which effectively suppress coupling surface currents between closely placed radiators.

Another mechanism proposed previously to isolate highly-coupled monopole antenna elements is by using 180^0 hybrid couplers [167–169]. The method is based on the mode-decomposition network, in which a multi-port network is inserted between the antennas and their driving ports.

To eliminate the use of extra resonating elements for enhancing isolation, use of a neutralization line between antenna elements is another promising technique. A neutralization-line cancels induced current with the direct current, since both currents have opposite phase [170–174]. However, the electrical length of the naturalization line is a critical parameter and frequency dependent (inversely proportional).

Using a parasitic element between two closely placed antennas of a MIMO system, significant enhancement in isolation (low coupling) can be achieved [175–177]. The parasitic resonator basically induces a secondary coupling current of opposite phase that suppresses the direct coupling current between radiating elements. This also decreases the overall radiation efficiency of the system.

Another way to achieve high isolation between antenna elements in a compact MIMO system is to excite different modes having overlapping resonance. The different modes have different surface current distribution that minimize the coupling between the modes [178–180].

Chapter 3
UWB Antennas with Notch: Design and Performance Analysis

3.1 Introduction

In this chapter, UWB antenna with integrated band-notch functionality is presented and discussed. In the first section, design and analysis of a UWB antenna with single band-notch is presented. Tapering of the ground plane of a conventional printed monopole antenna is investigated to achieve wideband impedance matching and a split ring is used thereafter to realize the band-notch behavior. In second section, a printed UWB antenna is presented with a dual-band notch behavior at two distinct interference bands. In third section, a printed UWB antenna with tunable band-notch characteristic is presented. The notch band can be tuned electronically in real time to reject a particular frequency band. Antenna design, simulation and both frequency-domain and time-domain results of all the antennas are presented in subsequent sections.

3.2 Design and Analysis of Single Band-Notched Ultra-Wideband Antenna

3.2.1 Introduction and Related Work

Ultra-wideband communications have received much attention and been rapidly growing since the opening of 7.5 GHz frequency spectrum (3.1–10.6 GHz) because of its many inherent advantages such as high data rate, consumption of less average power, higher accuracy in range measurement and resolution, low system complexity and cost. Enhancing technologies demand for improvements like miniaturization, and cost reduction. UWB communications possess many challenges on the design of antennas because of pre-allocated spectrum. UWB systems are prone to interference from the existing narrow band systems such as IEEE 802.11a WLAN (5.15–5.35 GHz and 5.725–5.875 GHz) and HIPERLAN/2 (5.450–5.725 GHz), etc. This demands for a band-rejection in UWB transmission at these interfering bands. The popular techniques to realize band rejection characteristics include embedding slots or slits

J. Malik et al., *Compact Antennas for High Data Rate Communication*,
Springer Topics in Signal Processing 14, DOI 10.1007/978-3-319-63175-2_3

on the patch [71], placing parasitic elements near or at the rear end of the antenna [72], connecting quarter wavelength stubs to the patch [73, 74], connecting stubs to the feed-line [75] and placing resonators near the feed-line [76].

3.2.2 *Antenna Design and Implementation*

Suitable tapering the ground plane of a conventional monopole antenna results in a wide impedance bandwidth. As the first step to design the ground plane for the proposed UWB antenna, a non-linear taper with given end diameters and length is designed for a raised cosine shape profile 3.1. Using in-house code, this taper is designed and further optimized using particle swarm optimization (PSO) for the maximum mode coupling and enhanced operating frequency for a relatively small taper length. A detailed procedure/algorithm to design the specific taper with desired performance can be found in [77]. The ground plane of the present UWB antenna is raised cosine tapered, which enhances coupling between modes that ensures good impedance match over a broad frequency range. The synthesis of the raised cosine taper profile is carried out using the following expressions given in [77]. The taper profile between r_1 and r_2 can be computed as (3.3) and is shown in Fig. 3.1.

$$\alpha = -1.0 + 2.0(\frac{i}{l})^{\gamma} \tag{3.1}$$

$$r(y) = \frac{r_2 - r_1}{2}(\alpha + \frac{1}{\alpha}\sin(\pi\alpha)) + \frac{r_2 - r_1}{2} \tag{3.2}$$

$$r = r_1 + r(y) \tag{3.3}$$

where, γ is the geometrical parameter with value 0.94. For the particular taper design, the values are $r_1 = \mathrm{L1}$, $r_2 = \mathrm{L}$, and $l = \mathrm{W}$. Figure 3.1b shows the geometry of the proposed UWB monopole antenna. The antenna is fabricated on FR-4 substrate (ε_r = 4.4) with thickness of 1.524 mm. The circular radiating element with multi-section feeding line is printed on one side of the substrate and the tapered ground plane on the other side. Stepped impedance transformer like transmission line is used as feed-line to the antenna. The length and width of various sections of feed-line are optimized parametrically during simulation for improved impedance matching (Fig. 3.1). To achieve the characteristic line impedance of 50 Ω at the port, width of the microstrip feed-line that is soldered to the SMA connector is chosen accordingly. Figure 3.1c shows the antenna impedance bandwidth with different feed-line dimensions. With the optimized feed-line, the antenna shows wideband impedance matching. The design parameters of the antenna are given in Table 3.1.

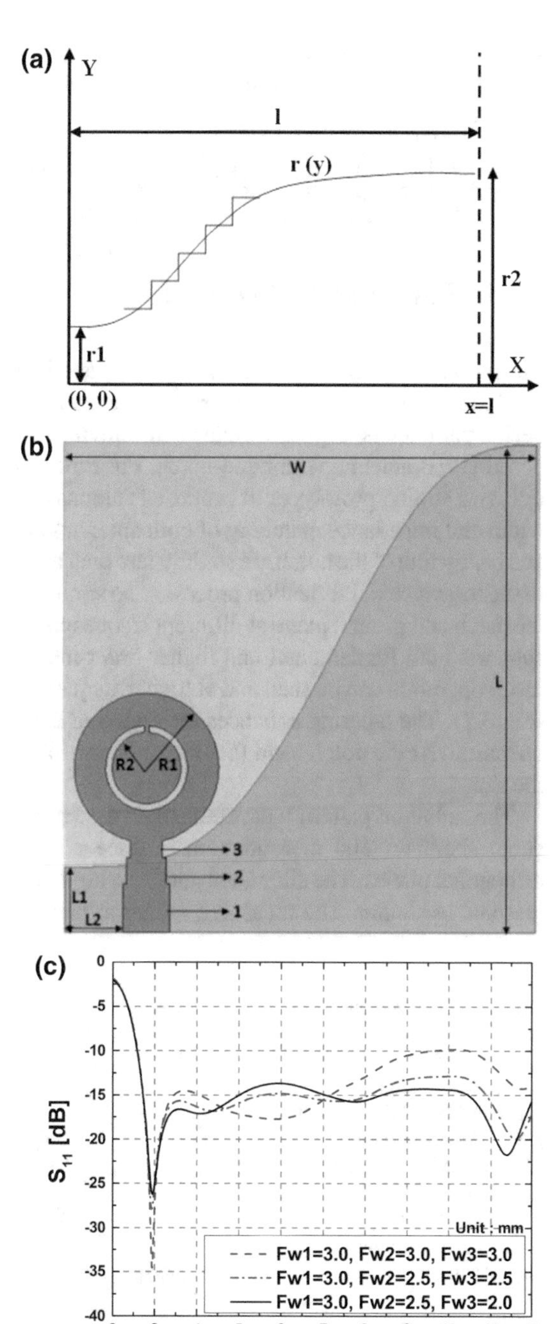

Fig. 3.1 **a** Raised-cosine taper **b** band-notched UWB antenna with tapered ground plane **c** simulated return loss with different feed configurations without notch [181] [From: Jagannath Malik and M.V. Kartikeyan, "Band-notched UWB antenna with raised cosine-tapered ground plane," Microwave and Optical Technology Letters, vol. 56, no. 11, pp. 2576–2579, 2014. Reproduced courtesy of the John Wiley & Sons, Ltd.]

Table 3.1 Design parameters and values of single band-notched UWB Antenna [181] [From: Jagannath Malik and M.V. Kartikeyan, "Band-notched UWB antenna with raised cosine-tapered ground plane," Microwave and Optical Technology Letters, vol. 56, no. 11, pp. 2576–2579, 2014. Reproduced courtesy of the John Wiley & Sons, Ltd.]

Paramater	L	W	L1	L2	R1	R2	Fw1	FL1	Fw2	FL2	Fw3	FL3
Value (mm)	30	30	4.0	3.7	4.6	2.2	3.0	3.0	2.5	1.6	2.0	1.4

3.2.3 Results and Discussion

The proposed antenna is designed and simulated in finite integration technique based CST Microwave studio v12. A parametric study is carried out to optimize various design parameters. Figure 3.2 shows the simulated and measured return loss of proposed UWB antenna with band-notch. For time domain performance characterization, two similar prototypes of proposed antenna are fabricated. The measured bandwidth and impedance matching of both antennas are similar. The measured strength and bandwidth of the notch are slightly less than the simulated which can be attributed to the tolerance in fabrication process. The simulated surface current distribution on the patch and ground plane at different frequencies is shown in Fig. 3.2. The antenna supports both fundamental and higher order modes. At 3.0 GHz, the fundamental mode operation can be seen and at higher frequency higher harmonics are generated (Fig. 3.2). The tapering enhances the coupling between modes resulting wide band operation. At the notch band (5.8 GHz), the surface current is confined only around the slot.

The radiation pattern measurement of the fabricated antenna is done inside anechoic chamber. The measurement is done at three different frequencies in two orthogonal planes. The simulated and measured radiation patterns of the antenna are reasonably similar. The measured radiation patterns are shown in Fig. 3.3. The H-

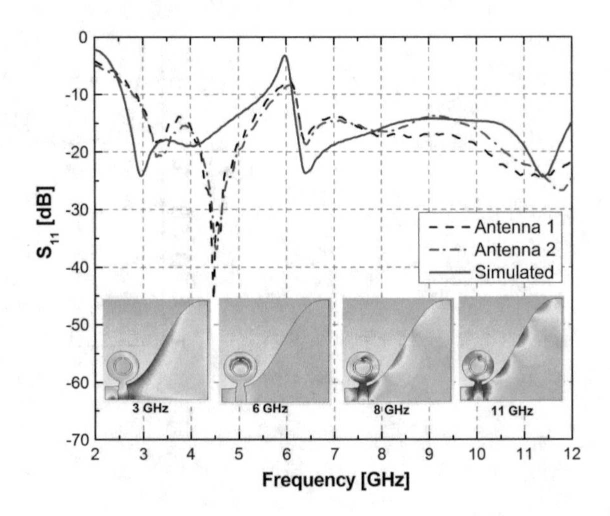

Fig. 3.2 Simulated and measured return loss of fabricated antenna; surface current distribution at different frequencies in inset [181] [From: Jagannath Malik and M.V. Kartikeyan, "Band-notched UWB antenna with raised cosine-tapered ground plane," Microwave and Optical Technology Letters, vol. 56, no. 11, pp. 2576–2579, 2014. Reproduced courtesy of the John Wiley & Sons, Ltd.]

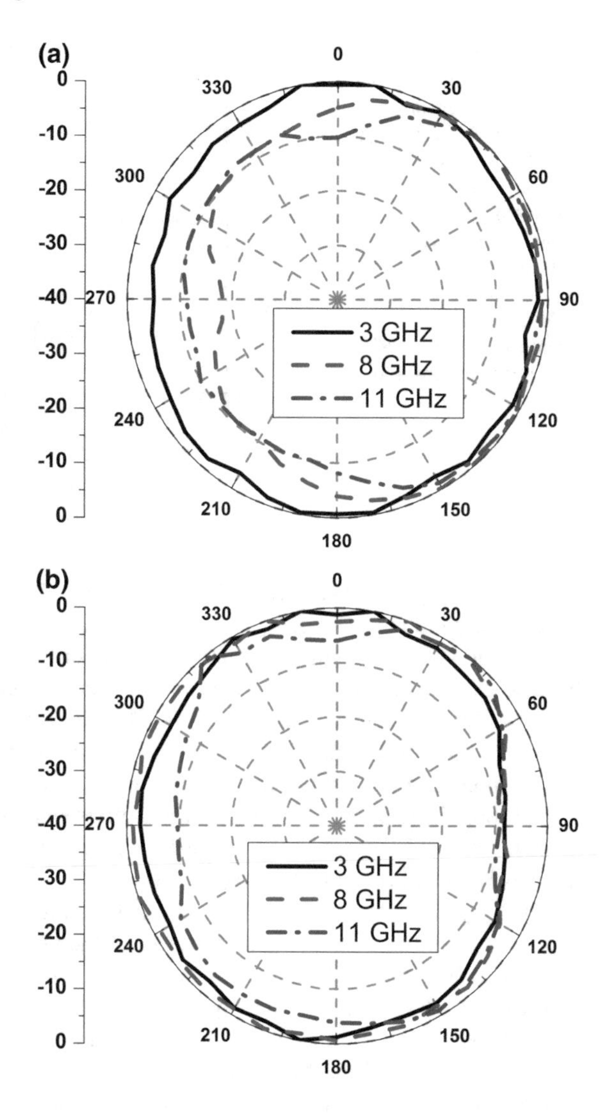

Fig. 3.3 Measured radiation ratterns of UWB antenna with notch in **a** E–plane (Y–Z) and **b** H–Plane (X–Z) [181] [From: Jagannath Malik and M.V. Kartikeyan, "Band-notched UWB antenna with raised cosine-tapered ground plane," Microwave and Optical Technology Letters, vol. 56, no. 11, pp. 2576–2579, 2014. Reproduced courtesy of the John Wiley & Sons, Ltd.]

Plane patterns are fairly omni-directional in the entire bandwidth. The gain measurement is done against a standard gain horn antenna using substitution/transfer method. The maximum measured gain of the UWB antenna with notch is around 4.1 dBi at 11 GHz. Figure 3.4 shows measured gain Vs frequency plot. The fabricated antenna is shown in inset. The dip in the gain is very narrow at the notch frequency which can be observed in Fig. 3.4. The measured gain is less than 5 dBi over the entire bandwidth.

The antenna system measurement is carried outside the anechoic chamber considering a real world situation where multiple reflecting objects are present. The antenna system involves two identical band-notched UWB antennas kept at a

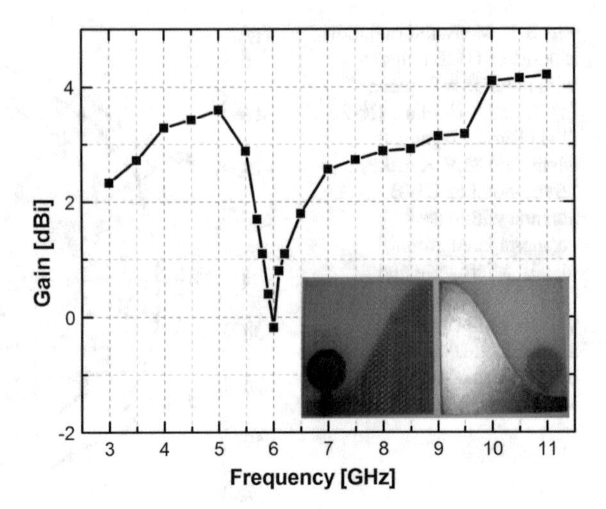

Fig. 3.4 Fabricated antennas and measured gain [181] [From: Jagannath Malik and M.V. Kartikeyan, "Band-notched UWB antenna with raised cosine-tapered ground plane," Microwave and Optical Technology Letters, vol. 56, no. 11, pp. 2576–2579, 2014. Reproduced courtesy of the John Wiley & Sons, Ltd.]

distance of 30 cm between them, in line of sight (LOS) scenario. Both antennas are connected to the ports of VNA, one being transmitter and other as receiver. The investigation of the system performance is done for two different cases. In first case, the antennas are oriented face-to-face to each other. In the second case, antennas are kept side-by-side to each other (Fig. 3.5). The transmission characteristics for both cases are shown in Fig. 3.5. In the S_{21} parameter of the band-notched UWB antenna system, there is a sharp dip at notch frequency for both cases. The variation in measured group delay of the UWB antenna system is small over the entire band except at the notch-band. There is a sharp change in the group delay at notch frequency due to the presence of phase nonlinearities (Fig. 3.5).

The S_{21} parameter was imported to the MATLAB for further analysis. It is assumed that the transmitter transmits 4th derivative of the Gaussian pulse and the received pulse is to be computed. The particular signal is chosen since it has permissible power spectal density as specified by the FCC mask for indoor communications [78]. The detailed procedure to find out the received pulse shape and SFF is explained in Chap. 2. The SFF reaches unity as the two pulses are exactly same in shape, which means that the antenna system does not distort the incident pulse at all. The normalized transmitted signal and normalized received signal is shown in Fig. 3.6. The received signal in face-to-face and in side-by-side case is shown in Fig. 3.6a, b respectively. Although the distortion in received signal is higher in second case compared to first case, still its acceptable as the SFF for both cases are higher than 0.5. The SFF for present band-notched UWB antenna system is 0.902 and 0.814 for first and second case respectively.

The ringing duration is calculated as the time required for the signal to attenuate from maximum value of the signal to 5% of the maximum value of the signal. The ringing durations of the received signal for band-notched UWB antenna system in

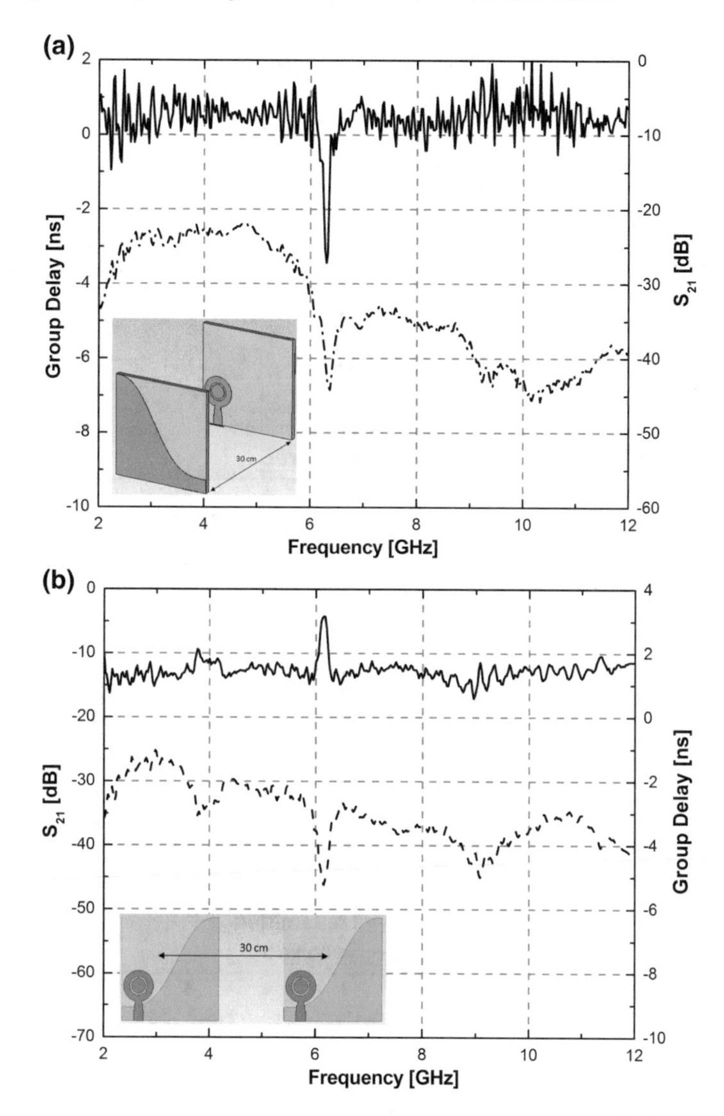

Fig. 3.5 Measured system transfer function and group delay **a** face-to-face **b** side-by-side orientations (*solid line*: group delay, *dashed line*: S_{21}) [181] [From: Jagannath Malik and M.V. Kartikeyan, "Band-notched UWB antenna with raised cosine-tapered ground plane," Microwave and Optical Technology Letters, vol. 56, no. 11, pp. 2576–2579, 2014. Reproduced courtesy of the John Wiley & Sons, Ltd.]

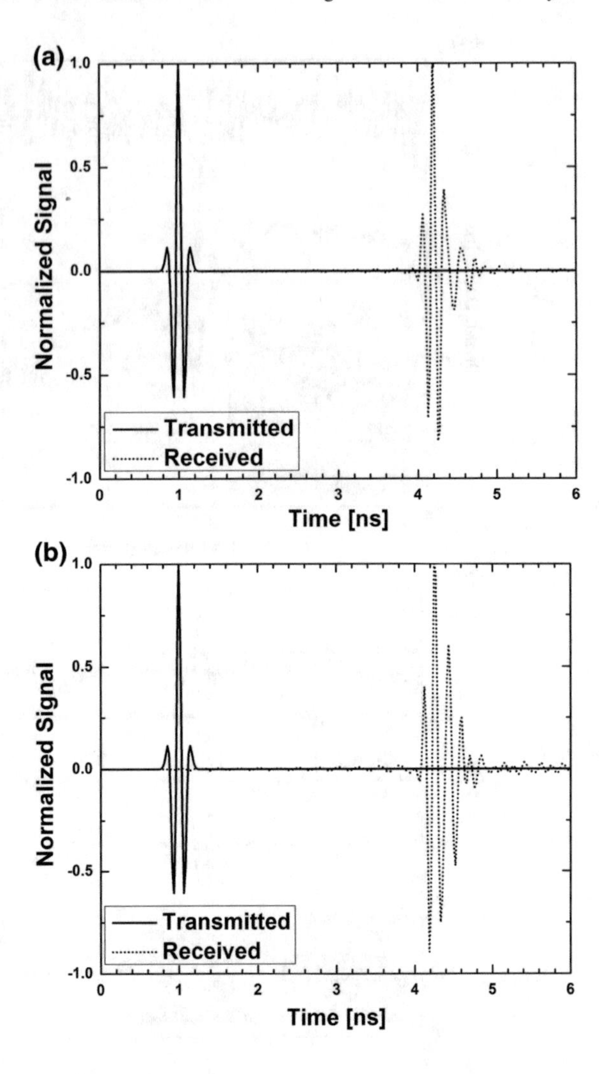

Fig. 3.6 Normalized received signals for band-notched UWB antenna system in **a** face-to-face **b** side-by-side orientation [181] [From: Jagannath Malik and M.V. Kartikeyan, "Band-notched UWB antenna with raised cosine-tapered ground plane," Microwave and Optical Technology Letters, vol. 56, no. 11, pp. 2576–2579, 2014. Reproduced courtesy of the John Wiley & Sons, Ltd.]

both orientations are significantly small (Fig. 3.6). The received signal shows less distortion and fairly retains the transmitted pulse shape. This shows the received signal is less susceptible to Inter Symbol Interference (ISI).

3.3 Design and Analysis of Dual Band-Notched Ultra-Wideband Antenna

3.3.1 Introduction and Related Work

Due to large bandwidth, UWB systems may experience interference from other popular narrow band communication systems like wireless local area network (WLAN) (5.15–5.35 GHz, 5.725–5.825 GHz) and world interoperability for microwave access (WiMAX) (3.4–3.69 GHz, 5.25–5.825 GHz) [80]. There are two possible solutions to reduce interference with these systems. Those are either to design a source pulse that has zero PSD at above mentioned bands, or to design band-notched UWB antenna that does not radiate at interfering bands. Various band-notched techniques are reported in [81–86], such as half wave length slots on radiator or ground, $\lambda_g/4$ length open ended slot on patch (V-shape, arc-shape, U-shape, H-shape and F-shape) and use of parasitic elements. To create multiple notches, parasitic loading technique requires multiple resonators and the electrical length of each parasitic resonators needs to be $\lambda_g/2$ length at their respective notch frequencies. This is a major drawback considering compactness of antenna. Moreover, large conductor size and electromagnetic coupling between these parasitic resonators add to the worst.

To overcome these limitations, dual band-notched antenna is proposed. A CPW feed resonator and a C-shape slot on the radiator are used to get the dual band-notch behavior. Operation of CPW resonator is independent of the size or shape of radiating patch. During designing of CPW resonator, notch frequency is precisely modified/tuned just by changing the physical dimension of the resonator. CPW feeding is used because of better performance at higher frequency, simplicity and cost effective due to single side metalization.

3.3.2 Antenna Design and Implementation

The geometry of proposed UWB monopole antenna with dual band-notched characteristics is shown in Fig. 3.7. The antenna is implemented on a FR–4 substrate with dielectric constant as 4.4 and thickness as 1.524 mm. The width of feed-line that is soldered to the SMA connector and the gap between feed-line and ground is chosen to achieve the characteristic line impedance of 50 Ω at the connecting port. Rectangular bevel cut is analyzed with the help of full-wave EM simulator to enhance impedance bandwidth. Simulated bandwidth of CPW fed antenna without any stepped cut and antenna with cuts at lower side of the patch and upper side of the ground is shown in Fig. 3.7b. The antenna labeled shown in Fig. 3.7b is the initial CPW fed antenna which is matched in the lower band of the desired bandwidth. Beveling the lower corner of the patch with a single rectangular shape cut (labeled) improves the matching at higher band and increases the impedance bandwidth as shown in Fig. 3.7b. Final design with rectangular cuts at lower corner of the patch

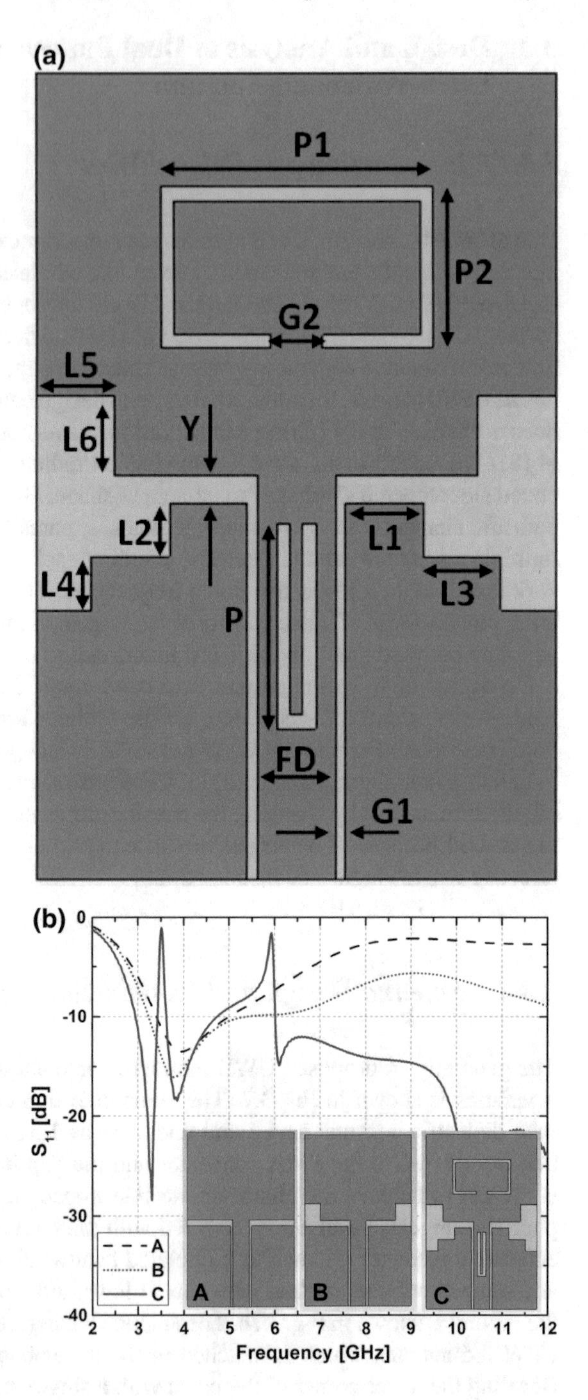

Fig. 3.7 **a** Schematic of proposed dual band-notch antenna **b** simulated return loss [182] [From: Jagannath Malik, Parth C. Kalaria and M.V. Kartikeyan, "Transient response of dual-band-notched ultra-wideband antenna," *International Journal of Microwave and Wireless Technologies* vol. 7, no. 1, pp. 61–67, 2014. Reprinted with permission of the EuMA/Cambridge University Press]

Table 3.2 Design parameters and values of dual band-notched UWB antenna [182] [From: Jagannath Malik, Parth C. Kalaria and M.V. Kartikeyan, "Transient response of dual-band-notched ultra-wideband antenna," *International Journal of Microwave and Wireless Technologies* vol. 7, no. 1, pp. 61–67, 2014. Reprinted with permission of the EuMA/Cambridge University Press]

Paramater	FD	G1	Y	P	L1	L2	L3	L4	L5	L6	P1	P2	G2
Value (mm)	3	0.4	1.1	7.4	3	3	3	2	3	3	10.5	6	2

and the upper corner of the ground shows a wideband matching from 3.1–10.6 GHz. Parametric analysis is adopted to optimize the dimensions of these cuts to improve matching over the wide band frequency.

To realize the band-notch two slots are added to the reference UWB antenna. A slot at the upper portion of the antenna is created, that creates a notch at lower 3.5 GHz band. A CPW resonator is used to get the second band-notch at 5–6 GHz band. The antenna geometry is unaffected by changing center frequency of rejected band at 3.5 GHz and/or at 5–6 GHz band. The level of impedance matching for both reference UWB antenna and antenna with notch is similar except at the dual notch-bands. The final optimized antenna with dual band-notch is shown in Fig. 3.7a and the simulated return loss in Fig. 3.7b labeled as 'C'. The design parameters and its dimensions are given in Table 3.2.

3.3.3 Results and Discussion

The center frequency of both notches can be tuned by changing the physical dimensions of the slots. Figure 3.8 shows parametric analysis of slot length on their corresponding notch frequencies. The notch frequency obeys the inverse relationship with the slot length. Increasing slot lengths increase the current patch, hence the notch frequency moves to the lower side (Fig. 3.8). It can be observed that tuning a particular notch does not affect the other notch frequency. Hence coupling between slot and CPW resonator is very less and both can be tuned independently. Figure 3.8a, b show the parametric analysis of CPW resonator length 'P' and slot parameter 'P1' respectively. It an observation that the bandwidth of notch bands remains unaffected to the variation in lengths. Tunning of the notch band has a little effect on the impedance bandwidth except at notch band.

To have a deeper insight on the band-notch behavior of antenna, surface current distribution is analyzed and shown in Fig. 3.9 for 3.5 and 5.8 GHz. At 3.5 GHz, the surface current is concentrated around the upper slot, and at 5.8 GHz, it distributed around the CPW resonator. The coupling between the resonators is minimum.

To measure the time-domain performance of proposed dual band-notched antenna, two identical antennas are fabricated. Figure 3.10 shows fabricated antennas and the measured return loss for both antennas. The measured return loss of both the fabricated antennas is quite similar and in accordance to the simulated return loss.

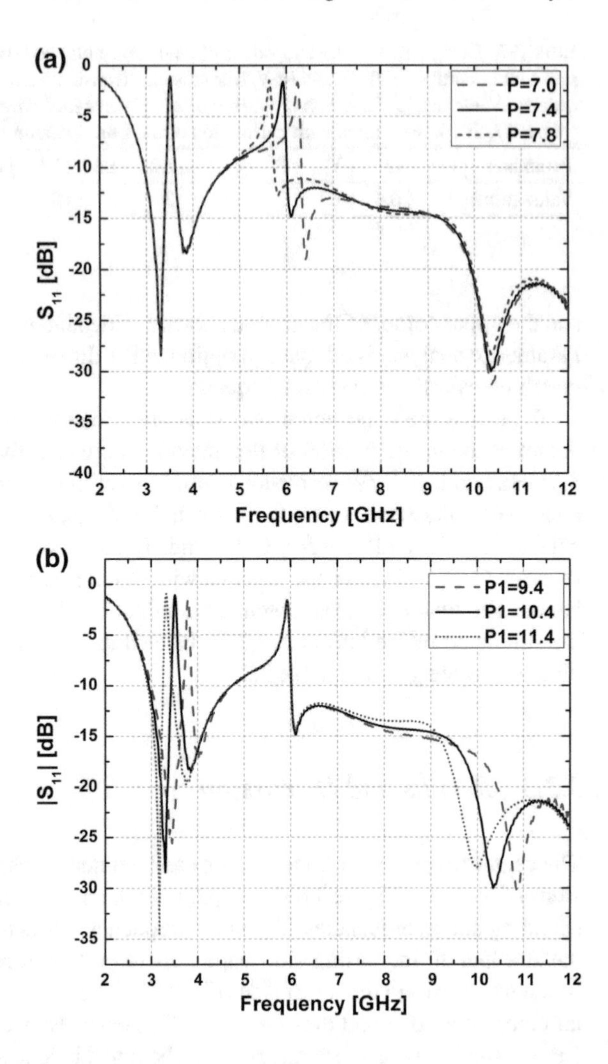

Fig. 3.8 Effect of parametric variations on notch frequency for proposed antenna [182] [From: Jagannath Malik, Parth C. Kalaria and M.V. Kartikeyan, "Transient response of dual-band-notched ultra-wideband antenna," *International Journal of Microwave and Wireless Technologies* vol. 7, no. 1, pp. 61–67, 2014. Reprinted with permission of the EuMA/Cambridge University Press]

The fabricated antennas successfully operate with an extended impedance bandwidth (measured at 10 dB return loss) covering the FCC recommended 3.1–10.6 GHz band. Dual band-notch is achieved exactly at 3.5 and 5–6 GHz, but the measured notch strength is little less than the simulated strength. The variation between measured and simulated results are within acceptable limits.

It is desired that the far–field radiation pattern of an UWB antenna should be omni-directional and stable over the entire impedance bandwidth. The omni-directional nature of the pattern limits the realized gain compared to a directional antenna. The measured gain of the fabricated antenna is shown in Fig. 3.10b. Since both the fabricated antennas are identical, gain plot of single antenna has been shown. The gain is measured at discrete frequencies along the direction of maximum power reception using standard gain horn antenna DRH10 (amitec) with substitution/transfer method.

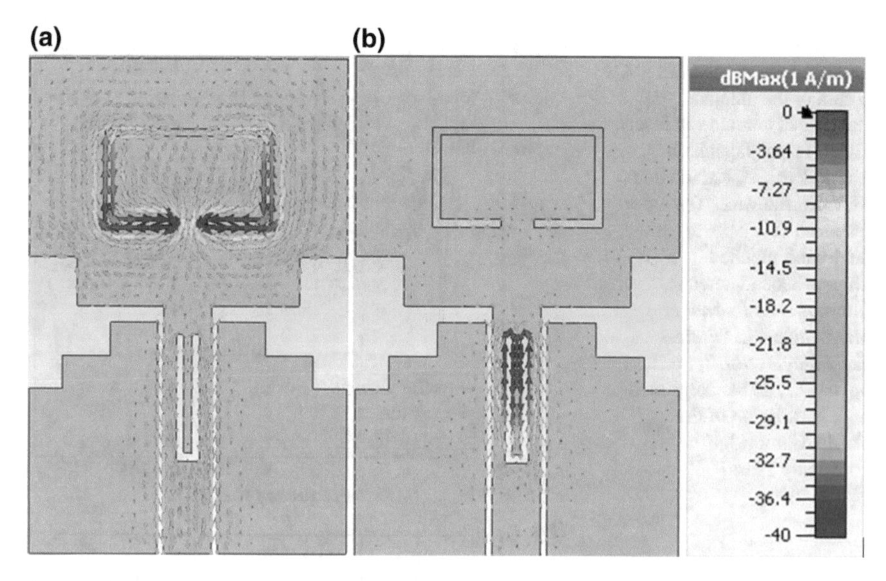

Fig. 3.9 Simulated surface current distribution at **a** 3.5 GHz **b** 5.8 GHz [182] [From: Jagannath Malik, Parth C. Kalaria and M.V. Kartikeyan, "Transient response of dual-band-notched ultra-wideband antenna," *International Journal of Microwave and Wireless Technologies* vol. 7, no. 1, pp. 61–67, 2014. Reprinted with permission of the EuMA/Cambridge University Press]

Measured gain has a flat nature over the frequencies except at the notch-bands. The dip in the gain is very narrow at the notch frequency which can be observed from Fig. 3.10. The measured gain is less than 5 dBi over the entire bandwidth.

The normalized radiation patterns of the band-notched UWB antenna measured inside anechoic chamber at different frequencies for E and H-planes are shown in Fig. 3.11. The measured power patterns in H-Plane are fairly omnidirectional over the entire band. It shows the doughnut shaped pattern at 4 GHz, which is similar to a conventional monopole in free space. The E-Plane power patterns are dumbbell shaped at the lower frequency and changes at higher band due to higher order mode propagation.

The transmission characteristics of the antenna system is measured outside anechoic chamber considering the real world situation. The antenna system involves two identical band-notched UWB antennas, one as transmitting and other as receiving antenna. The antennas are separated by a distance of 50 cm in their LOS. The investigation of the system performance is carried out in two different orientations. The antennas are oriented face-to-face in one measurement, and side-by-side to each other in another measurement (Fig. 3.12). The group delay variation of UWB antenna system is <1 ns in the entire band for both orientations. In the S_{21}-parameter of band-notched UWB antenna system, there is a sharp dip at notch frequency. There is a sharp change in the group delay at the notch frequency for band-notched UWB antenna system because of the presence of phase nonlinearities.

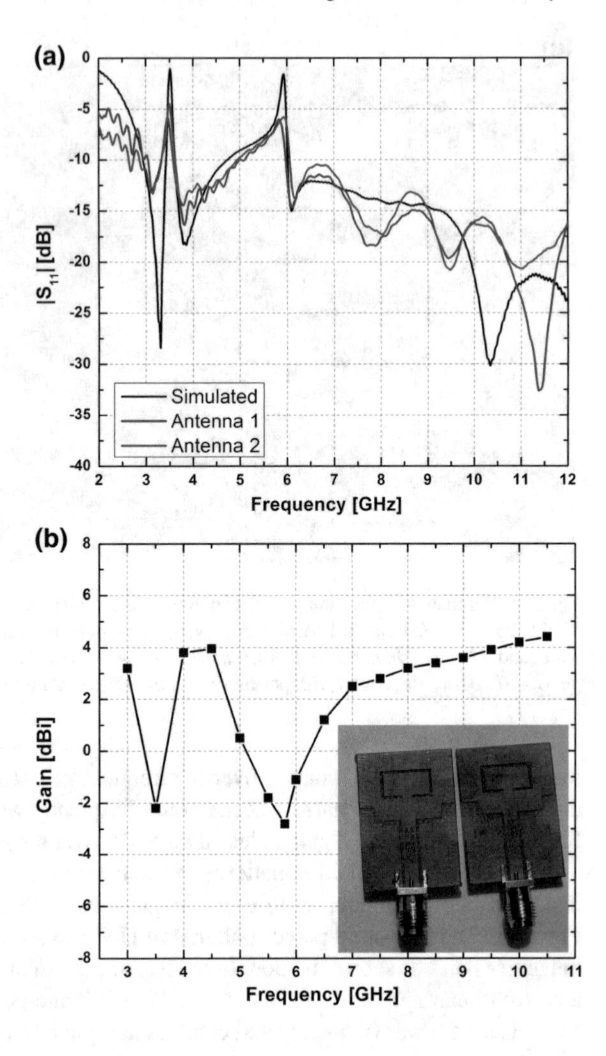

Fig. 3.10 Measured **a** return-loss of antennas **b** gain of the antenna (fabricated antennas in inset) [182] [From: Jagannath Malik, Parth C. Kalaria and M.V. Kartikeyan, "Transient response of dual-band-notched ultra-wideband antenna," *International Journal of Microwave and Wireless Technologies* vol. 7, no. 1, pp. 61–67, 2014. Reprinted with permission of the EuMA/Cambridge University Press]

The antennas are assumed to be excited by a 4th derivative of the Gaussian pulse UWB signal which satisfies the FCC power spectral mask for indoor communication [78]. The system fidelity factor (SFF) is defined as the cross-correlation between the normalized input signal $T_s(\mathrm{t})$ and normalized received signal $R_s(\mathrm{t})$ [79]. The detailed procedure to extract received pulse from system transmission characteristics (S_{21}) is explained in Chap. 2. Figure 3.13 shows the normalized input signal and normalized received signal for present dual band-notched UWB antenna system. The SFF for band-notched UWB antenna system in face-to-face and side-by-side orientations are 0.821 and 0.807 respectively. The ringing durations in received signal is small.

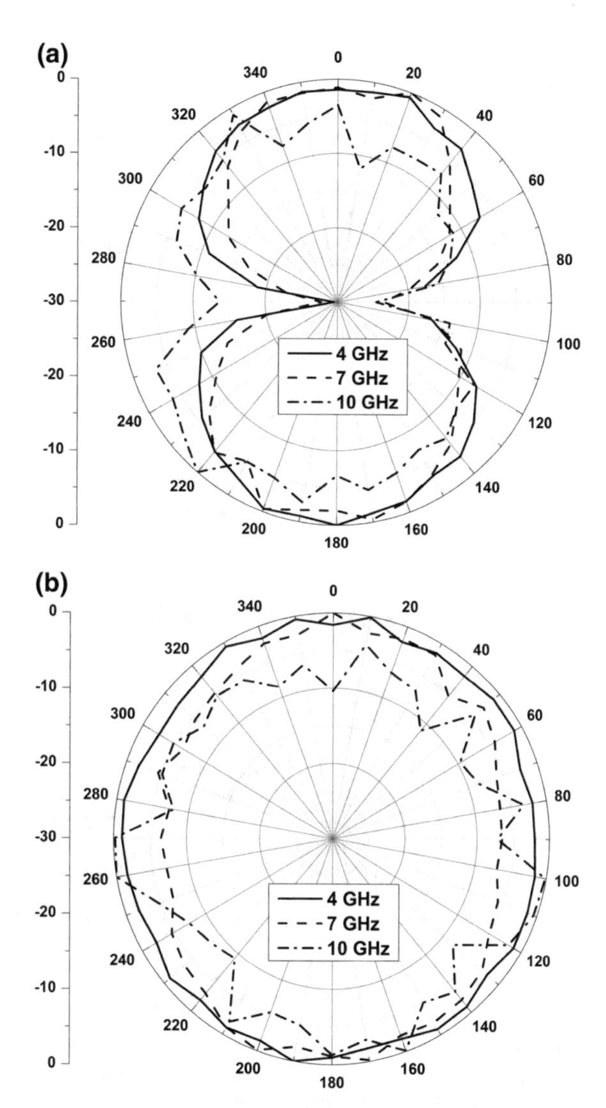

Fig. 3.11 Measured radiation pattern at different frequencies in **a** E–plane, **b** H–plane [182] [From: Jagannath Malik, Parth C. Kalaria and M.V. Kartikeyan, "Transient response of dual-band-notched ultra-wideband antenna," *International Journal of Microwave and Wireless Technologies* vol. 7, no. 1, pp. 61–67, 2014. Reprinted with permission of the EuMA/Cambridge University Press]

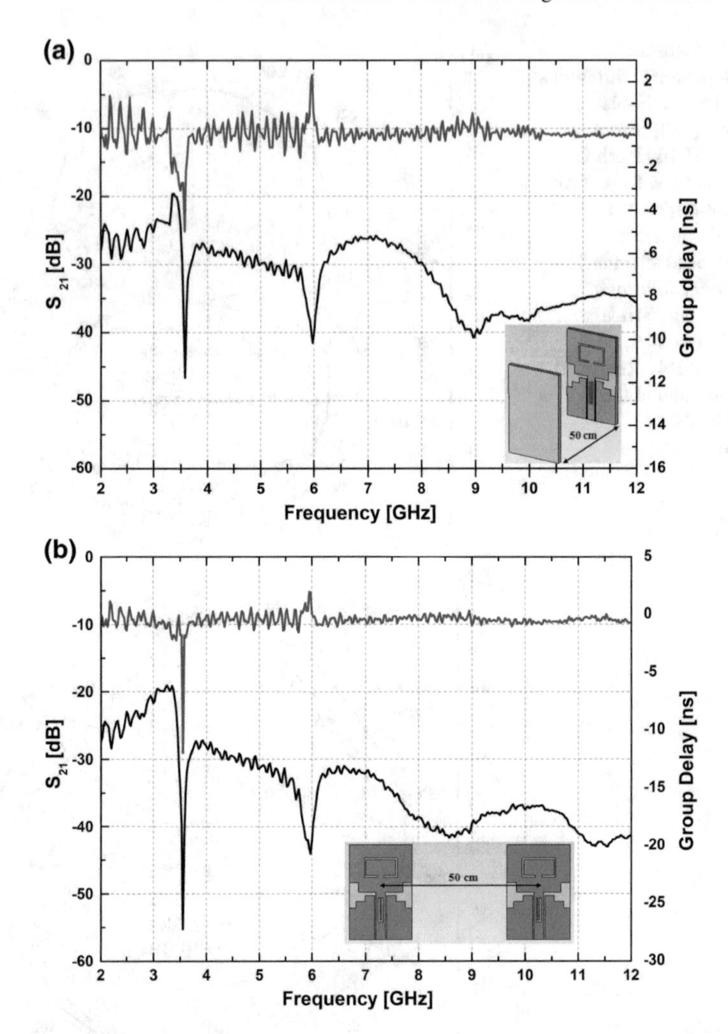

Fig. 3.12 Measured transmission and group delay in **a** face-to-face **b** side-by-side orientations (*black line*, S_{21}; *red line*, group delay) [182] [From: Jagannath Malik, Parth C. Kalaria and M.V. Kartikeyan, "Transient response of dual-band-notched ultra-wideband antenna," *International Journal of Microwave and Wireless Technologies* vol. 7, no. 1, pp. 61–67, 2014. Reprinted with permission of the EuMA/Cambridge University Press]

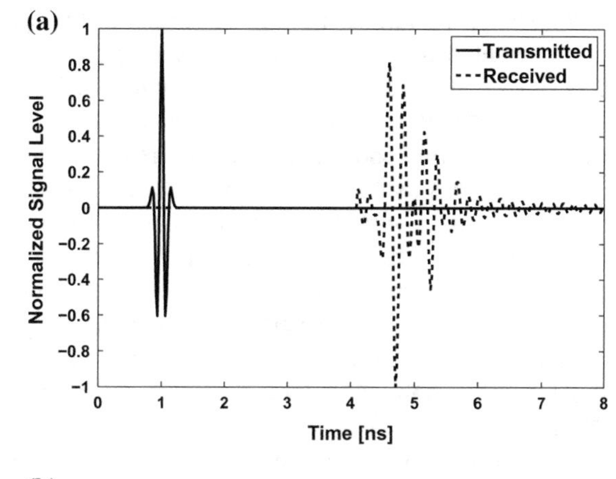

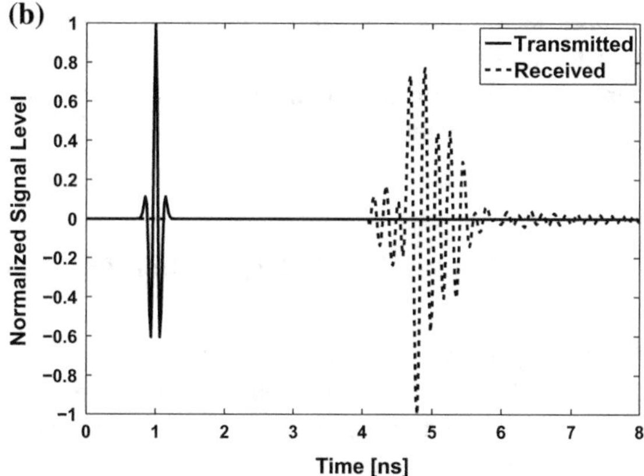

Fig. 3.13 Normalized received signals **a** face-to-face and **b** side-by-side orientation [From: Jagannath Malik, Parth C. Kalaria and M.V. Kartikeyan, "Transient response of dual-band-notched ultra-wideband antenna," *International Journal of Microwave and Wireless Technologies* vol. 7, no. 1, pp. 61–67, 2014. Reprinted with permission of the EuMA/Cambridge University Press]

3.4 Design and Analysis of Electronically Tunable Band-Notched Ultra-Wideband Antenna

3.4.1 Introduction and Related Work

Due to huge operating bandwidth, UWB systems are often questioned as an issue of possible interference to some existing popular narrow band communication systems e.g. IEEE 802.11a WLAN (5.15–5.35 GHz and 5.725–5.875 GHz) and HIPERLAN/2 (5.450–5.725 GHz), etc. UWB antennas integrated with inherent band-rejection capability are preferred over antenna with external RF filters. Some techniques to create band-notch are using square-ring resonator [87], EBG structures [88], embedding slots on patch, ground [89] and feed-line [90], using parasitic resonator, quarter wavelength stubs connected to patch, stubs connected to the feed line, and placing resonators near feed line [73, 87–93] etc.

Typically, the interfering signals vary with the spatiotemporal scenario of the environment. Most of the techniques as mentioned above gives a static band-notch behavior, i.e. the notch band can't be altered once the antenna is fabricated. Flexibility/tunability in terms of rejected band is necessary so as to reduce the interference from systems which operate in different frequency bands. Varactor diodes are good candidate in this manner, since the capacitance of varactor diode can be varied by changing the bias voltage. If varactor diode is used in notch design, the tuning of rejected band can be done with a proper biasing circuit. Generally loops or stubs are loaded with varactor diodes to tune the notched band [94, 150]. These methods require more space on the antenna, thereby decreasing the possibility of multi-notch designs. Moreover, all these designs make use of two varactor diodes to produce a single notch and the time domain characteristics of the designed antennas are not discussed.

In the following section, we will discuss a compact CPW-fed UWB antenna with tunable narrow notched band characteristic. A quarter wavelength stub loaded with varactor diode facilitates continuous tunable band-notch function. The frequency domain and time domain characteristics of the fabricated band-notched UWB antenna system are presented.

3.4.2 Antenna Design and Implementation

As first step to design the tunable band-notched UWB antenna, a CPW fed elliptically tapered monopole antenna is designed and optimized to resonate over 3.1 to 10.6 GHz. The antenna is realized on a FR–4 substrate (ε_r = 4.4, tan δ = 0.0024, h = 1.524). The overall footprint of the antenna is ($L \times W$, L = 30 mm, W= 24 mm). Elliptical tapering of the patch and ground plane adjacent to the center metal in the CPW feed, and trimming of the ground below the radiating patch are the techniques used to enhance the impedance bandwidth. The study of various parameters is carried out to investigate the performance of antenna with respect to the effect on impedance bandwidth. A quarter wavelength stub is used on the rear side of the antenna and shorted to the CPW feed line through a metallic via of 0.6 mm diameter. The geometry of the proposed compact UWB antenna that produces a fixed notch at 5.8 GHz is shown in Fig. 3.14. The position at which the stub has been shorted to the feed line is an important parameter which affects both notch strength and bandwidth. Both notch strength and bandwidth decrease with the stub moving closer to the patch ('vh' increasing), as shown in Fig. 3.15. The antenna design parameters are given in Table 3.3.

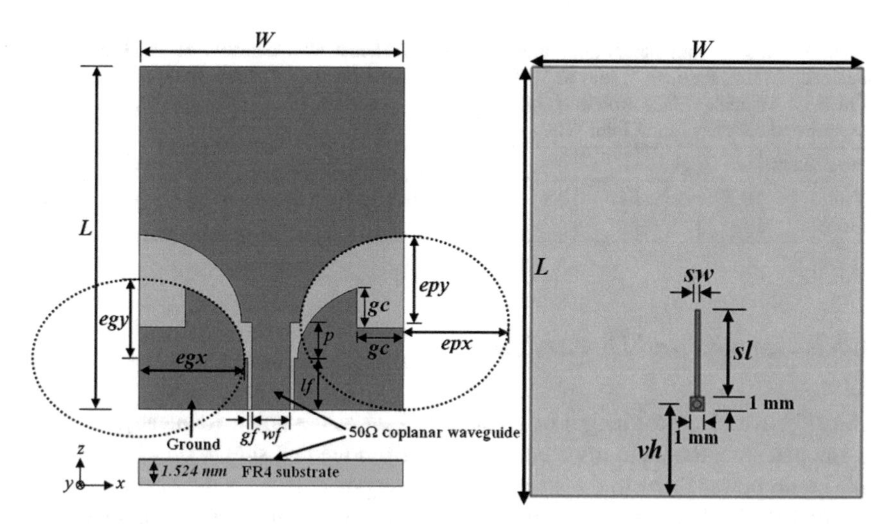

Fig. 3.14 Schematic view of CPW-fed band-notched UWB antenna [183] [From: Jagannath Malik, Paritosh V. and M.V. Kartikeyan, "Continuously Tunable Band-notched Ultra-Wideband Antenna," *Microwave and Optical Technology Letters*, vol. 57, no. 4, pp. 924–928, 2015. Reproduced courtesy of the John Wiley & Sons, Ltd.]

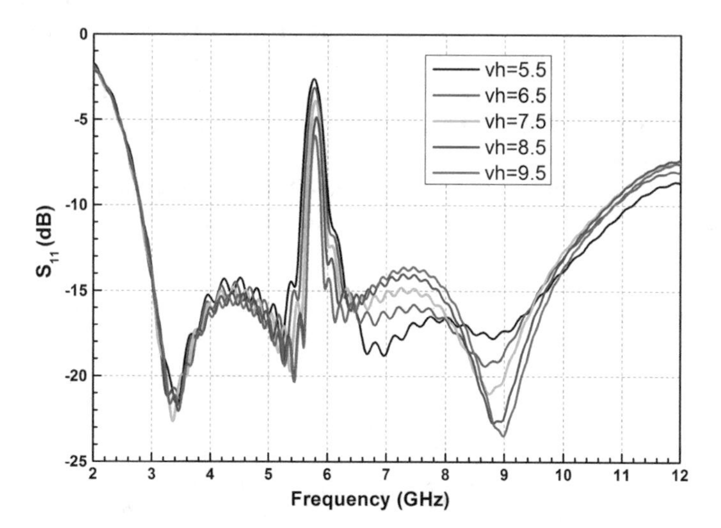

Fig. 3.15 Parametric analysis of 'vh' (other dimensions are kept fixed) [183] [From: Jagannath Malik, Paritosh V. and M.V. Kartikeyan, "Continuously Tunable Band-notched Ultra-Wideband Antenna," *Microwave and Optical Technology Letters*, vol. 57, no. 4, pp. 924–928, 2015. Reproduced courtesy of the John Wiley & Sons, Ltd.]

Table 3.3 Design parameters and values of UWB Antenna with tuneable notch [183] [From: Jagannath Malik, Paritosh V. and M.V. Kartikeyan, "Continuously Tunable Band-notched Ultra-Wideband Antenna," *Microwave and Optical Technology Letters*, vol. 57, no. 4, pp. 924–928, 2015. Reproduced courtesy of the John Wiley & Sons, Ltd.]

Parameters	wf	gf	lf	p	epx	epy	egx	egy	gc	vh	sl	sw
Values (mm)	0.39	0.38	4.5	3.1	9.32	7.66	9.57	6.85	4.18	6.5	6.1	0.32

3.4.3 Results and Discussion

The effective electrical length of the stub is $\lambda_g/4$ at the notch frequency. Since, it is an open stub, the input impedance as seen from feeding side is very low. Hence, maximum power flows to the stub and very less power reaches the patch. This can be confirmed through the simulated surface current distribution on the antenna at notch frequency, and also through the simulated real part of the input impedance of the antenna as shown in Fig. 3.16.

In order to tune notched band, the stub is loaded with a Skyworks SMV 1232 varactor diode as shown in Fig. 3.17. The total capacitance of the Varactor diode can be varied from 0.72 pF at 15 V bias to 4.15 pF at 0 V bias voltage. A DC block capacitor of 2.2 pF is used to block the DC. An inductor of 10 nH is used to isolate the RF from DC bias. A resistor (1.5 Ω), an inductor (0.7 nH) and a variable capacitor (0.72–4.15 pF) are connected in series to form an equivalent circuit for varactor diode in the CST simulation.

The measured return loss of band-notched UWB antenna loaded with $\lambda_g/4$ stub and varactor diode at different reverse bias voltage is shown in Fig. 3.18. With an

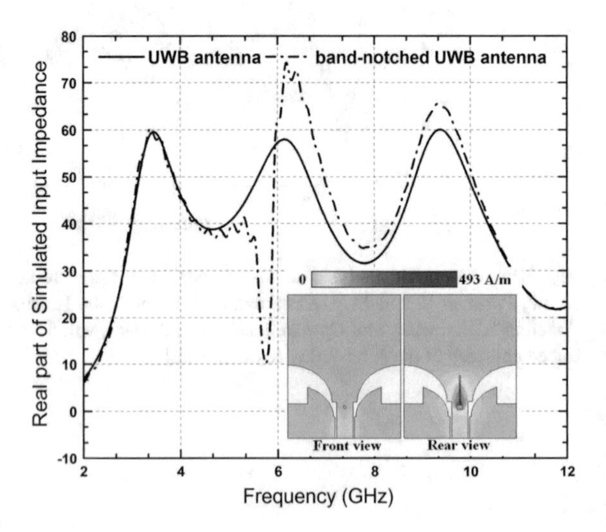

Fig. 3.16 Simulated real part of input impedance and surface current at notched-band [183] [From: Jagannath Malik, Paritosh V. and M.V. Kartikeyan, "Continuously Tunable Band-notched Ultra-Wideband Antenna," *Microwave and Optical Technology Letters*, vol. 57, no. 4, pp. 924–928, 2015. Reproduced courtesy of the John Wiley & Sons, Ltd.]

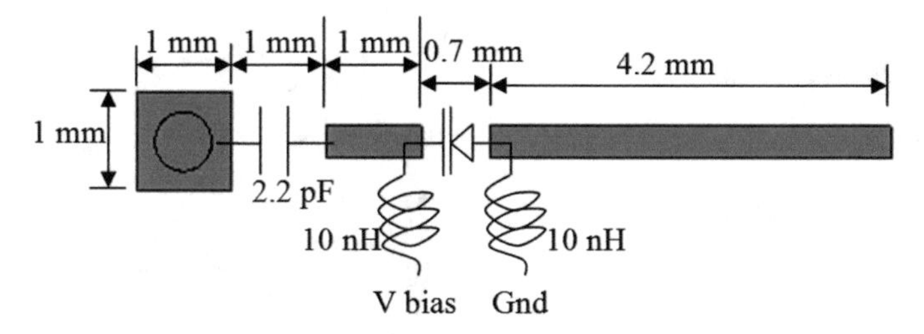

Fig. 3.17 Schematic of stub loaded with varactor diode [183] [From: Jagannath Malik, Paritosh V. and M.V. Kartikeyan, "Continuously Tunable Band-notched Ultra-Wideband Antenna," *Microwave and Optical Technology Letters*, vol. 57, no. 4, pp. 924–928, 2015. Reproduced courtesy of the John Wiley & Sons, Ltd.]

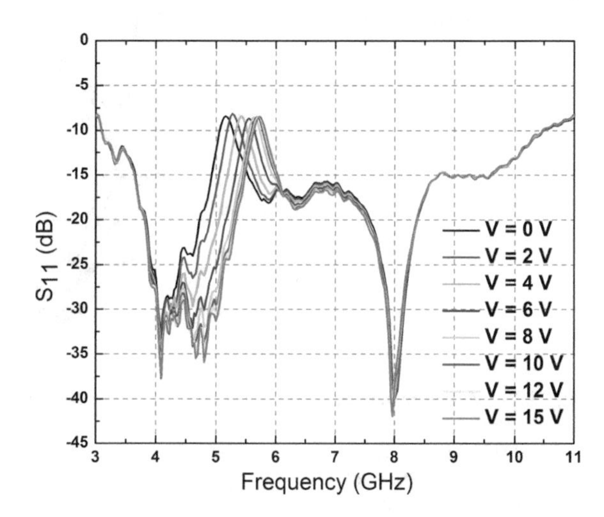

Fig. 3.18 Measured return loss of tunable notch UWB antenna with different bias voltages [183] [From: Jagannath Malik, Paritosh V. and M.V. Kartikeyan, "Continuously Tunable Band-notched Ultra-Wideband Antenna," *Microwave and Optical Technology Letters*, vol. 57, no. 4, pp. 924–928, 2015. Reproduced courtesy of the John Wiley & Sons, Ltd.]

applied reverse bias voltage to varactor diode varying from 0 V to 15V, the notched band can be tuned from (5.07–5.85) GHz with the center frequency varying from (5.1625–5.7125) GHz. The measured bandwidth of the notched-band at any reverse bias voltage is 0.195 GHz. The impedance bandwidth of the tunable notch UWB antenna is 7.42 GHz extending from 3.08 to 10.5 GHz measured at 10 dB return loss.

The radiation pattern and gain measurement is done at different frequencies inside anechoic chamber. The radiation patterns of UWB antenna with tunable notch at 0 V dc bias applied voltage are shown in Fig. 3.19. The H-Plane patterns are fairly omni-directional and E-Plane patterns are donut shaped in nature. The radiation pattern is consistent over the entire band. Figure 3.20 shows the fabricated antenna and the measured gain Vs frequency. The maximum measured gain of UWB antenna with tunable notch is 3.85 dBi. The dip in the gain is very sharp and narrow at the notch

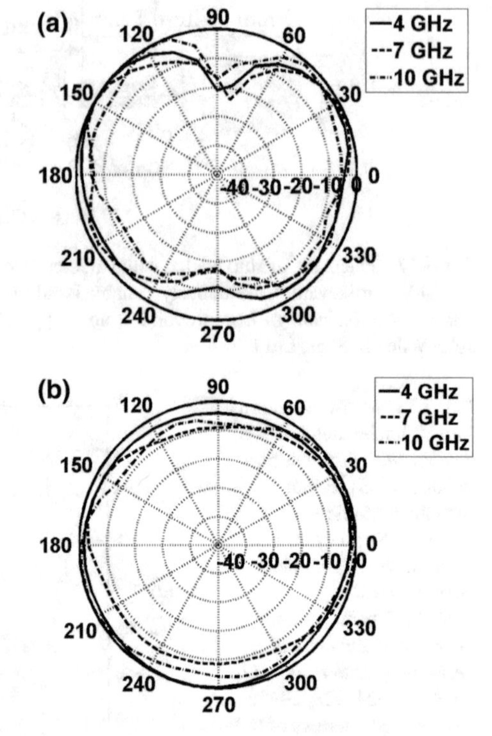

Fig. 3.19 Measured radiation patterns of UWB antenna with tunable notch at 0 V bias voltage in **a** E–plane (YZ-plane) and **b** H–plane (XZ-plane) [183] [From: Jagannath Malik, Paritosh V. and M.V. Kartikeyan, "Continuously Tunable Band-notched Ultra-Wideband Antenna," *Microwave and Optical Technology Letters*, vol. 57, no. 4, pp. 924–928, 2015. Reproduced courtesy of the John Wiley & Sons, Ltd.]

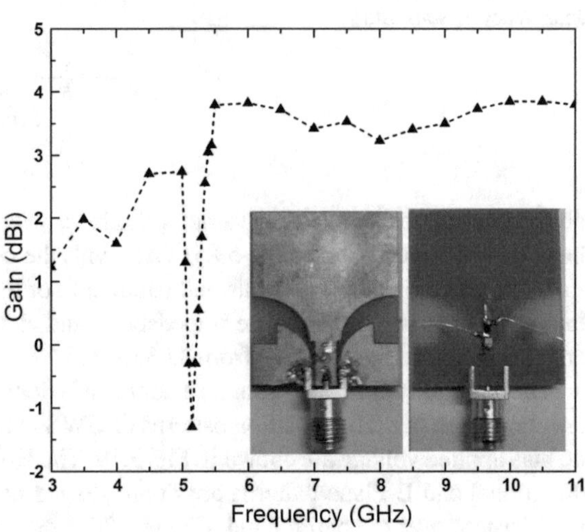

Fig. 3.20 Measured gain of UWB antenna with tunable notch at 0 V bias voltage (fabricated antenna in inset) [183] [From: Jagannath Malik, Paritosh V. and M.V. Kartikeyan, "Continuously Tunable Band-notched Ultra-Wideband Antenna," *Microwave and Optical Technology Letters*, vol. 57, no. 4, pp. 924–928, 2015. Reproduced courtesy of the John Wiley & Sons, Ltd.]

frequency which can be observed from Fig. 3.20. The measured gain flatness over the impedance bandwidth can be observed with peak gain less that 5 dBi over the entire bandwidth.

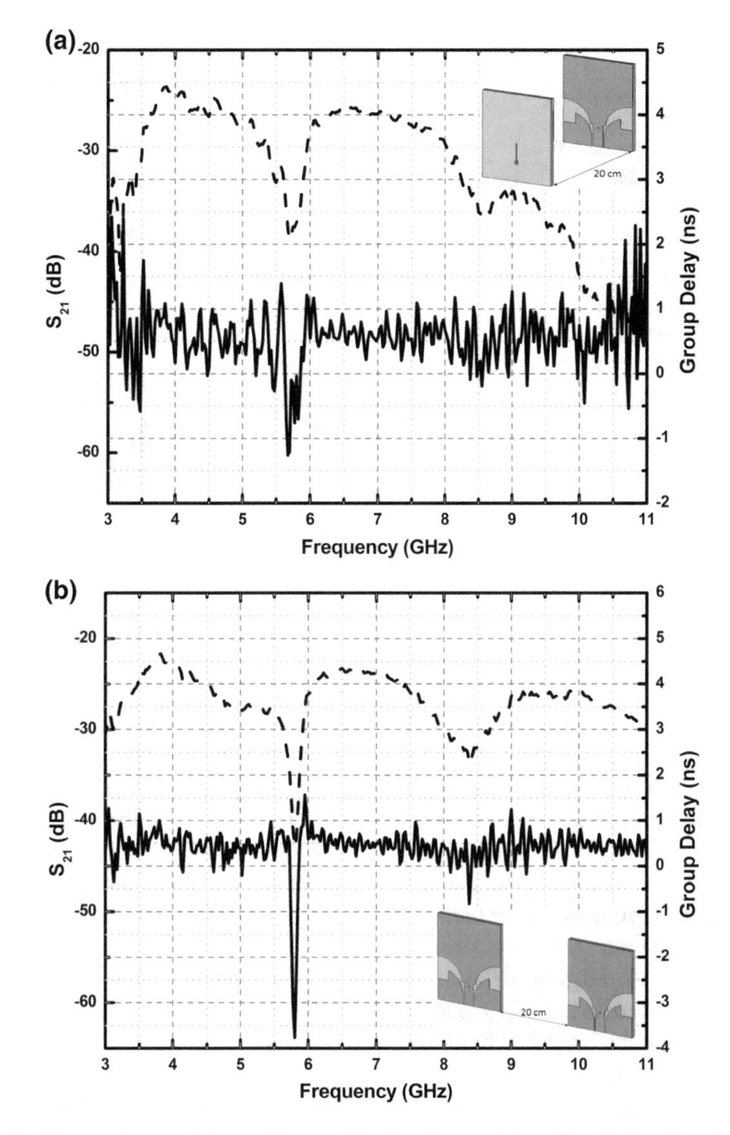

Fig. 3.21 Measured transmission and group delay in **a** face-to-face and **b** side-by-side orientations (S_{21}: *dashed line*; group delay: *solid line*) [183] [From: Jagannath Malik, Paritosh V. and M.V. Kartikeyan, "Continuously Tunable Band-notched Ultra-Wideband Antenna," *Microwave and Optical Technology Letters*, vol. 57, no. 4, pp. 924–928, 2015. Reproduced courtesy of the John Wiley & Sons, Ltd.]

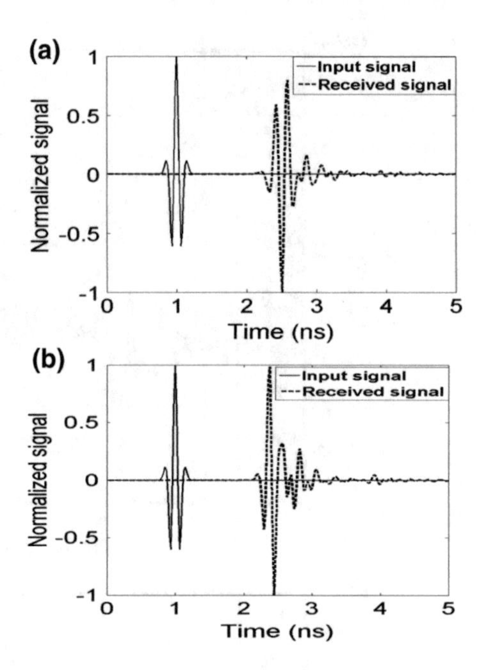

Fig. 3.22 Normalized received signals in **a** face-to-face **b** side-by-side orientations [183] [From: Jagannath Malik, Paritosh V. and M.V. Kartikeyan, "Continuously Tunable Band-notched Ultra-Wideband Antenna," *Microwave and Optical Technology Letters*, vol. 57, no. 4, pp. 924–928, 2015. Reproduced courtesy of the John Wiley & Sons, Ltd.]

The transmission characteristics of UWB antenna system is done outside anechoic chamber considering a real world situation. The antenna system involves two identical UWB antennas separated by a distance of 20 cm located in LOS (Fig. 3.21). The detailed procedure to extract received pulse from system transmission characteristics (S_{21}) is explained in Chap. 2. The group delay is an important parameter to characterize the linear response of a system. Figure 3.21 shows the measured S_{21}-parameter and group delay for both cases. In the S_{21}-parameter of the band-notched UWB antenna system, there is a sharp dip at 5.8 GHz for both the orientations. The variation in measured group delay of the band-notched antenna system is small except at the notch. There is a dip in group delay at 5.8 GHz (notch frequency) in both the orientations for the band-notched UWB antenna system because of the presence of phase nonlinearities.

Figure 3.22 shows the normalized transmitted and received signal for both the cases. The ringing durations in received signal in face-to-face and side-by-side orientations are very low. This shows that the received signal is less susceptible to ISI. The SFF for UWB antenna system in face-to-face and side-by-side orientations are 0.92 and 0.85 respectively.

3.5 Concluding Remarks

In this chapter, UWB antenna with single, dual and tunable band-notch behavior have been presented. In first and second designs, the realized band-notch is static, i.e. once the antenna is fabricated, the notch frequency can't be changed. In the first design, a printed circular UWB monopole antenna with raised-cosine tapered ground plane has been analyzed. The novelty of the antenna is tapered ground plane which enhances mode coupling to achieve wide impedance bandwidth. Since the ground plane is smooth and tapered, the localized field distribution is less compared to ground plane with sharp edges. To realize the notch band at 5.8 GHz, a SRR type slot is etched from the circular patch. In the second design, a CPW-fed UWB antenna has been designed with dual band-notch characteristics. Slot loaded radiator and 'U' shaped CPW resonators have been used for band rejection at 3.5 and 5–6 GHz respectively. In third design, UWB antenna with electronically tunable notch is discussed. A $\lambda_g/4$ stub loaded with a varactor diode has been used to tune the notched band. The notched bandwidth is narrow, highly selective, and continuously tunable with an applied reverse bias voltage to the varactor diode. Simulation and measurement results of all antennas in frequency and time domain have been presented in detail.

It is observed that various band-notch techniques behave all most similarly in the frequency domain. The notch bandwidth and strength can be tuned with parametric optimization. So to quantify the effectiveness of different band-notch techniques, a time domain analysis and comparison will be helpful. The same is addressed in Chap. 4.

Chapter 4
Band-Notch Techniques in UWB Antennas: A Comparison

4.1 Design and Analysis of Band-Notch Techniques in UWB Antenna

4.1.1 Introduction and Related Work

Ultra-wideband systems are popularly categorized as impulse radio (IR) communications, as in time domain, the data is transmitted in the form of extremely narrow pulses without any conventional modulation. Antennas play a critical role at the front-end of transceiver for successful implementation of UWB standards. UWB antennas should be capable of operating over bandwidth with reasonable efficiency and satisfactory radiation properties. At the same time it should posses a good time domain performance, i.e. a good impulse response with minimal distortion to transmitted pulse [109]. It is desired that the pulse should get detected at the receiver end with minimum distortion. Due to huge operating bandwidth, UWB systems are prone to interfere with existing narrow band communication systems. External band-stop filter can be used with UWB antennas to mitigate interference issue, but makes overall system bulky and adds non-linearity to the system. So, UWB antennas with integrated band-rejection capability are preferred [65–67]. These techniques can be categorized broadly into two types, (a) using resonant slots, (b) using parasitic resonators to the reference UWB antenna. In the former type, slots are either cutaway from the radiating element [110], embedded in the ground plane [111], or inserted in the feed-line itself [112]. The later type consists of parasitic resonators placed close to the feed line or in the vicinity of the radiating element [113]. The rejection band can be tuned by changing the dimensional parameters and location of these slots or parasitic resonators. All the above methods have their own pros and cons [108, 109].

When slots are used on the radiating element of the UWB antenna to create a notch, basically it creates destructive interference of the currents at a particular frequency. On the other hand, placing slots on the ground plane or on the feed line itself, behaves as a pre-filter that creates a rejection in the UWB impedance bandwidth.

J. Malik et al., *Compact Antennas for High Data Rate Communication*,
Springer Topics in Signal Processing 14, DOI 10.1007/978-3-319-63175-2_4

Using parasitic resonators also effectively creates a notch at the desired band. These techniques basically filter out the specific band from the UWB transmission, but it also adds some undesirable effects in time domain as well as in frequency domain behavior of the UWB system. In [108], authors illustrated a qualitative time domain performance comparison. It is reported that using slot on ground plane preserves transmitted the pulse shape with less ringing compared to slots on radiating patch. At the same, time these slots also disturbs the surface current distribution that may lead to deterioration of antenna radiation performance/pattern at other frequencies. These slots and parasitic elements behave as resonant type band-reject filters. These filter a specific spectral component and holds the energy, so that the UWB antenna ceases to transmit or receive at that frequency. However, these are low Q-filters and can't hold the energy for a longer time. Eventually, the stored energy decays in form of spurious radiation or losses. This affects the transient behavior of the antenna and causes prolonged ringing effect to the transmitted pulse in time domain.

Here, we present a comparative performance analysis (in frequency-domain and time-domain) of three popular band-notch techniques. The simulation results from CST Microwave studio v12 are presented with a special interest to the fidelity factor and ringing effect in time-domain due to band-notch techniques. Thereafter we present a band-notch method in a reference UWB antenna to realize a sharp rejection at WLAN band with less time domain ringing effect. The band rejection is achieved using a shorted quarter wavelength stub that effectively creates a very high impedance at the input port in the rejected band. A detail design methodology, simulated and measurement results are discussed in the subsequent sections.

4.1.2 Antenna Design and Band-Notch Implementation

The reference UWB antenna is simulated on FR–4 substrate ($\varepsilon_r = 4.4$, $\tan\delta = 0.0024$) with dielectric material thickness of 1.524 mm. The reference UWB antenna consists of an elliptical radiating element with multi-section feeding line on one side of the dielectric substrate, and the truncated ground plane on the other side of the substrate. The feed line is constructed as a stepped impedance transformer type with 3-section having different characteristic line impedance. Length and width of each sections of feed line was optimized parametrically for wide impedance matching. The upper part of the ground plane also plays an important role to achieve wideband operation at the lower resonance band. To further increase the impedance bandwidth, the edges of the ground plane were chamfered. The size of ground and the gap between elliptic patch and ground plane plays an important role to achieve wideband operation at the lower resonance band. To further increase the impedance bandwidth, the edges of the ground plane were chamfered.

Table 4.1 Design parameters and values for UWB antenna with 3-different notch techniques [184] [From: J. Malik, Amalendu Patnaik and M.V. Kartikeyan, "Time-domain Performance of Band-notch Techniques in UWB Antenna", *Proc. of the Asia-Pacific Microwave Conference (APMC-2016)*, IEEE, Dec. 2016. © IEEE 2016. Reproduced courtesy of the IEEE, USA.]

Paramater	Ex1	Ey1	C1	C2	C3	K1	K2	K3	K4	K5
Value (mm)	12	8.0	3.0	2.2	1.1	8.0	8.0	3.2	2.2	7.0
Parameter	K6	K7	K8	K9	K10	K11	K12	K13	G3	G4
Value (mm)	12.72	9.0	6.7	2.2	9.15	7.3	8.27	6.0	0.5	0.9

The first band-notch UWB antenna is realized by removing an annular split ring resonator (SRR) slot from the elliptical patch of the reference UWB antenna. Tuning the dimensions (length and width) of the slot and its location on the elliptic patch, the notch strength and notch bandwidth can be adjusted. However, the ground plane this antenna remains same as that of the reference UWB antenna. Figure 4.1a, b shows the top and bottom view of band-notch UWB antenna employing SRR slot.

The second band-notch UWB antenna is realized by placing a parasitically coupled resonator near to the feed-line of the reference UWB antenna. Figure 4.1c shows the second band-notch UWB antenna employing a rectangular split ring resonator which is parasitically coupled to the feed-line. The ground plane of this antenna is unaltered and similar to the ground plane of reference UWB antenna. The notch frequency and strength can be set by changing the total length of the split ring resonator and its distance from the feed-line respectively.

Figure 4.1d shows the third band-notch UWB antenna that employs a rectangular split ring slot on the ground plane of the reference UWB antenna. The top part of this antenna, i.e. the elliptic patch and feed-line are same as the reference UWB antenna. To get a symmetric ground current distribution and impedance profile, 2-slots are placed symmetrically w.r.t. the feed-line. The dimensions of all the design parameters have been given in Table 4.1.

The notch strength and notch frequency can be finely tuned by changing dimensional parameters and position w.r.t. the reference UWB antenna. With optimized design parameters, all the band-notch antennas cover the target FCC recommended spectrum with a band-notch behavior at 5.8 GHz to mitigate possible interference with WLAN communication. Figure 4.2 shows the simulated return loss of the reference UWB antenna and the three band-notch UWB antennas. It can be observed that all the band-notch antennas and reference UWB antenna have similar impedance matching profile except at the notch-band. The lower and higher cutoff frequencies (at return loss <10 dB) are same and unchanged by the notch techniques. This implies that the added parasitic resonator or slot only affects the intended notch band.

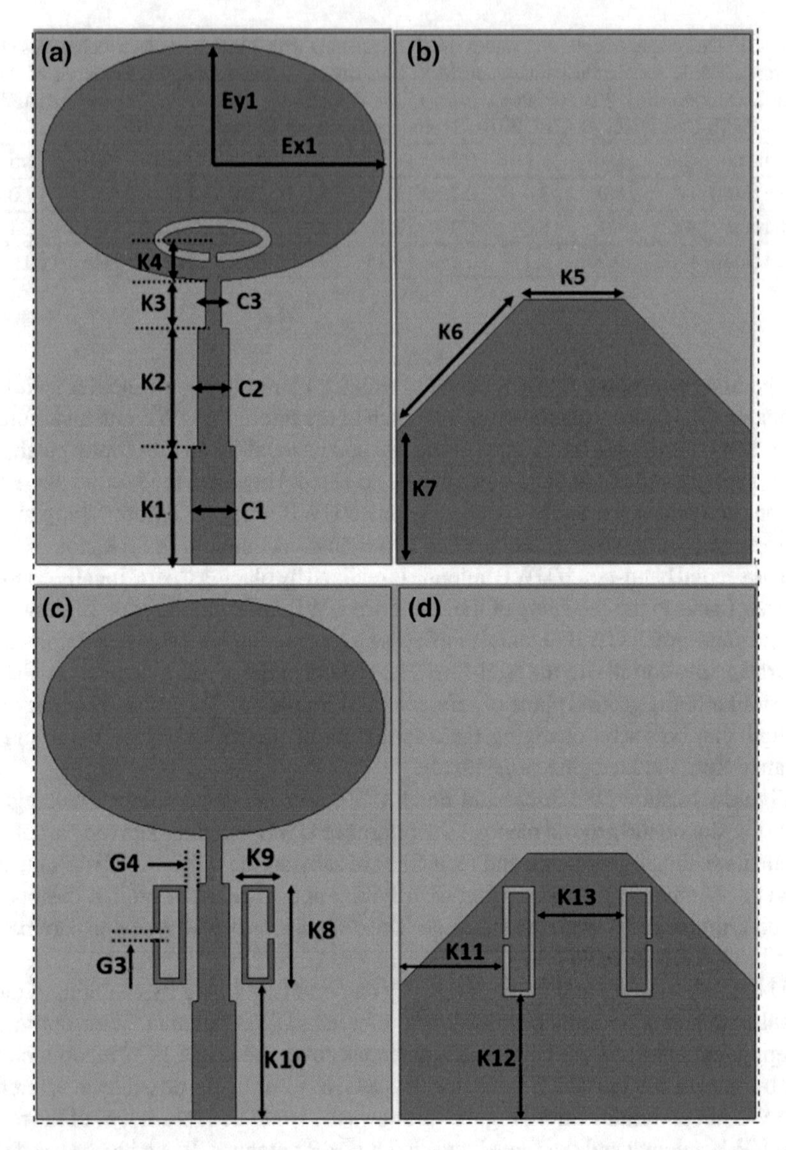

Fig. 4.1 Schematic of different band-notch techniques in reference UWB antenna [184] [From: J. Malik, Amalendu Patnaik and M.V. Kartikeyan, "Time-domain Performance of Band-notch Techniques in UWB Antenna", *Proc. of the Asia-Pacific Microwave Conference (APMC-2016), IEEE*, Dec. 2016. © IEEE 2016. Reproduced courtesy of the IEEE, USA.]

4.1.3 Time-Domain Performance Comparison of Band-Notch Techniques

From the simulated reflection coefficients as shown in Fig. 4.2, it can be clearly seen that the band-rejection strength of all the antennas is similar at the desired frequency of WLAN. This indicates that the three different notch-techniques considered for the present case, has similar frequency characteristics in terms of rejection strength and its bandwidth. For the time-domain performance characterization, the system transmission parameter for a pair of similar antenna located 50 cm apart in their line of site (LOS) have been analyzed in the simulation. The transmission characteristic (both magnitude and phase of S_{21}) are imported to the MATLAB for time-domain analysis. To investigate transmission parameter, 4th derivative of the Gaussian pulse is considered as the reference signal. The envelope of normalized received signal for all cases are shown in Fig. 4.3a, b, c and d.

The normalized transmitted and received signal shape for the reference UWB antenna without any band-notch is shown in Fig. 4.3a. It can be seen that the transmitted pulse goes negligible or almost zero distortion, which means the designed reference UWB antenna preserves the transmitted pulse shape at the receiver end. Assuming an ideal electromagnetic propagation environment without any kind of non-linearity, the designed reference UWB antenna is best suited for applications with low BER requirements. The normalized transmitted and received signal envelope for first, second and third band-notch antenna are shown in Fig. 4.3b, c and d respectively. It can be seen that the received signal undergoes significant ringing for all band-notch cases with different ringing magnitude and ringing duration in time domain. The techniques of using parasitic resonator near to the feed-line shows comparatively less magnitude of ringing and the ringing magnitude rapidly decaying over time. The method of using slot loaded ground plane shows higher ringing magnitude

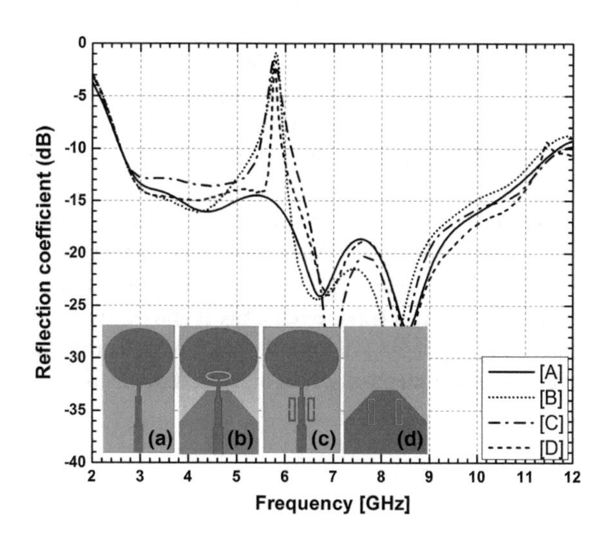

Fig. 4.2 Simulated S11 of reference UWB antenna-[A] and 3 band-notched antennas [B], [C],[D] [184] [From: J. Malik, Amalendu Patnaik and M.V. Kartikeyan, "Time-domain Performance of Band-notch Techniques in UWB Antenna", *Proc. of the Asia-Pacific Microwave Conference (APMC-2016), IEEE*, Dec. 2016. © IEEE 2016. Reproduced courtesy of the IEEE, USA.]

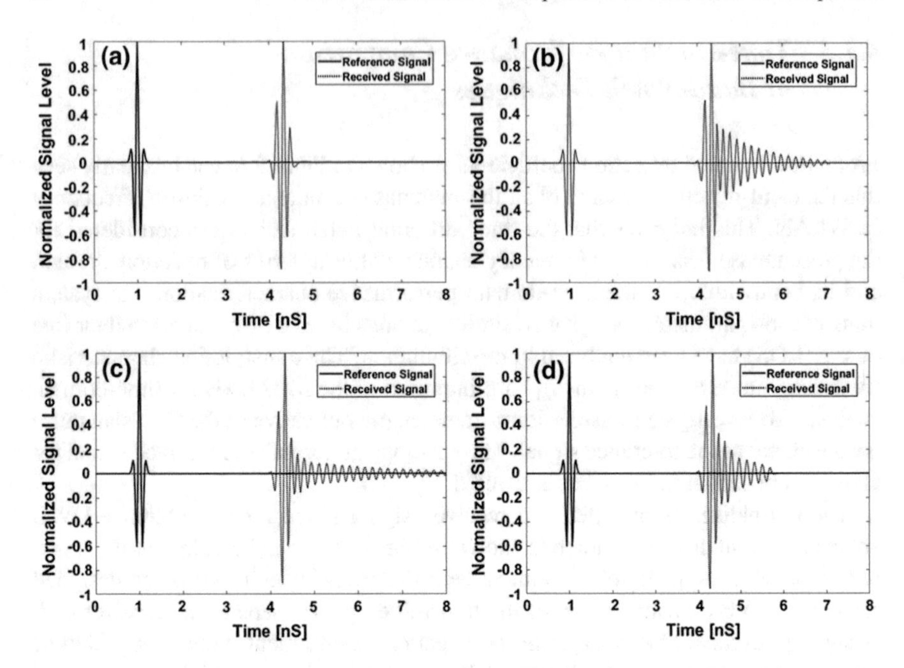

Fig. 4.3 Transmitted reference signal (*solid*); received signal (*dashed*) for **a** reference UWB antenna; antenna with **b** elliptic slot, **c** parasitic ring near feed-line, **d** slot on the ground plane [184] [From: J. Malik, Amalendu Patnaik and M.V. Kartikeyan, "Time-domain Performance of Band-notch Techniques in UWB Antenna", *Proc. of the Asia-Pacific Microwave Conference (APMC-2016), IEEE*, Dec. 2016. © IEEE 2016. Reproduced courtesy of the IEEE, USA.]

in the received pulse with a sustained oscillation over a long time. The technique of protuberating SRR from the radiating patch shows a similar behavior but with a less ringing duration. This ringing effect and pulse spreading are important factor to be considered when transmitting narrow pulses of the picosecond order. In high data rate communications the duty cycle of the carrier pulse is high. If the system undergoes pulse spreading and ringing, the probability of inter-symbol-interference (ISI) is high. For the present case it can be observed that the band-notch technique of using SRR slot gives a better time-domain performance compared to that of other two techniques. Since the surface current mainly confined to the edges, the effect of slots at frequencies other than the notch-band is minimum. The parasitic resonator filters a particular spectrum in the UWB, so that the antenna ceased to radiate at the notch frequency. Since this resonator is of low quality factor, it can't hold the filtered energy for a longer time. Eventually it ends up with radiating the filtered energy that results higher ringing in time-domain. A further investigation is being carried out to understand the mechanism along with the radiation properties.

Table 4.2 Design parameters and values of UWB antenna with $\lambda_g/4$ stub [185] [From: J. Malik, Amalendu Patnaik and M.V. Kartikeyan, "Novel band-notch technique for improved time domain performance of printed UWB antenna," *Proc. of URSI-RCRS 2015, India*. Reproduced courtesy of the URSI-RCRS, Belgium.]

Paramater	SW	SL	W1	W2	W3	L1	L2	L3	PW	PL	Y1	Y2
Value (mm)	20	30	3.0	2.0	1.0	6.0	6.0	4.0	4.0	7.8	13	10

4.2 Band-Notched UWB Antenna with Shorted Quarter-Wave Resonator

4.2.1 Antenna Design and Implementation

The geometry of the proposed elliptically tapered UWB antenna is shown in Fig. 4.4. The antenna is realized on FR-4 substrate with $\varepsilon_r = 4.4$ and dielectric thickness of 1.524 mm. The overall physical dimension of the antenna is (20 mm $\times$ 30 mm). Stepped impedance transformer like transmission line is used to feed the antenna. The feed-line is segmented into 3-sections with different widths for impedance matching over wide spectrum of UWB specifications. Selecting the number of sections and its dimensions (length and width) were intuitive, and optimized parametrically using CST Microwave studio v.12. The width of the microstrip feed-line that is soldered to the SMA connector is chosen as 3.0 mm to achieve the characteristic impedance of 50 Ω at the connecting port. With optimized feed-line, the antenna shows wide-band impedance matching. The upper section is elliptically tapered with lengths of minor axis and major axis as Rx2 = 12.0 mm and Ry2 = 17.5 mm respectively. Similarly, the lower part (ground plane) is also elliptically tapered with lengths of major axis and minor axis as Rx1 = 12.0 mm and Ry1 = 11.5 mm respectively. 'C2' and 'C1'are centers of the ellipses located at mid-point of the upper and lower edge of substrate as shown in Fig. 4.4. The extra portions of two ellipses extending outside the substrate edges has been removed. Parametric effect of the dimensions of two ellipses on impedance bandwidth is analyzed and optimized dimension is chosen accordingly. The feed-line is soldered to the lower end of upper ellipse using a through 'via'. At lower frequency the antenna behaves like resonator i.e. operating at fundamental mode. As we go towards the higher band, the antenna becomes traveling wave type with generation of higher order modes. Effective coupling between different resonances by means of elliptically tapered sections and matching due to stepped impedance feed–line gives ultra-wideband impedance matching response. The dimensions of design parameters are given in Table 4.2.

4.2.2 Results and Discussion

The simulated return loss for the reference UWB antenna and antenna with proposed notch technique are shown in Fig. 4.5. The level of impedance matching for both reference UWB antenna and band-notched UWB antenna is similar throughout the band, except at the notch–band at WLAN. The lower cutoff frequency of both antennas are same. The generalized expression to find out input impedance of a transmission line of length l with a termination load is given by:

$$Z_{in} = Z_0 \left[\frac{Z_L \cos \beta l + j Z_0 \sin \beta l}{Z_0 \cos \beta l + j Z_L \sin \beta l} \right] \tag{4.1}$$

where, Z_0 and Z_L are the characteristic and load impedance respectively with β as the propagation constant. The input impedance for a quarter wavelength line varies inversely with load impedance as given by:

$$Z_{in} = \frac{(Z_0)^2}{Z_L} \tag{4.2}$$

When quarter wavelength line is shorted with via, the input impedance will be high and behaves like an open circuit/band stop filter. The simulated input impedance plot for the proposed antenna from CST MWS is shown in Fig. 4.5. At notch frequency, the input impedance is high, as expected. The length of the quarter wavelength section 'PL' is calculated taking effective dielectric constant into account as given by:

$$\varepsilon_{eff} \approx \frac{\varepsilon_r + 1}{2} \tag{4.3}$$

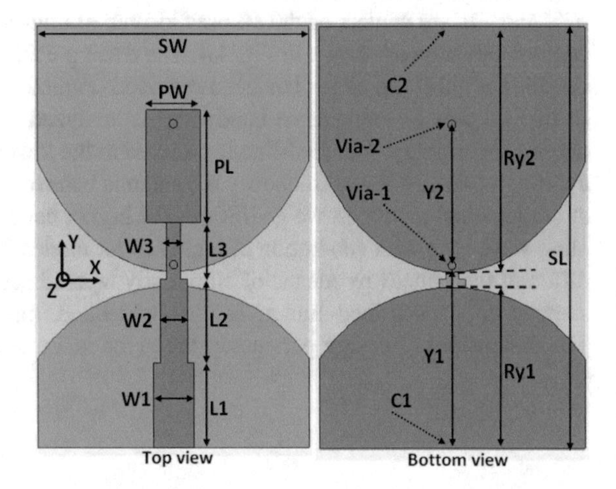

Fig. 4.4 Schematic of the proposed UWB antenna with $\lambda_g/4$ stub (*grey color*: metal) [185] [From: J. Malik, Amalendu Patnaik and M.V. Kartikeyan, "Novel band-notch technique for improved time domain performance of printed UWB antenna," *Proc. of URSI-RCRS 2015, India*. Reproduced courtesy of the URSI-RCRS, Belgium.]

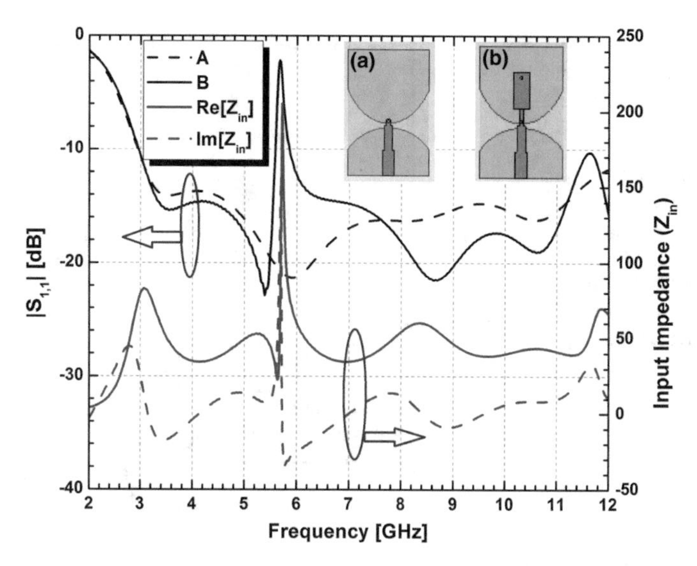

Fig. 4.5 Simulated return loss and input impedance for band-notched UWB antenna (*black color*: return loss; *blue color*: impedance) [185] [From: J. Malik, Amalendu Patnaik and M.V. Kartikeyan, "Novel band-notch technique for improved time domain performance of printed UWB antenna," *Proc. of URSI-RCRS 2015, India*. Reproduced courtesy of the URSI-RCRS, Belgium.]

At the notch frequency, it can be observed that the shorted stub acts like a resonant tank circuit which blocks current and behaves as open circuit (high impedance) at the input port. The real part of impedance is well around 50 Ω value throughout the operational band except at notch frequency. The imaginary part of impedance is close to zero except at notched band.

Figure 4.6 shows the parametric variation of return loss with length of shorted stub 'PL'. It can be seen that the notch frequency varies inversely with length of the stub, i.e. the notch can be tuned to desired frequency by controlling the stub length. The notch bandwidth is not affected by dimension 'PL', however the width 'PW' of the stub affects notch strength. The antenna with stub length 'PL' as 7.8 mm is fabricated and measured return loss is shown in Fig. 4.6.

The radiation pattern of the fabricated antenna is measured at discrete frequencies for both principal E and H-planes and shown in Figs. 4.7 and 4.8 respectively. The H-plane patterns looks fairly omnidirectional and the E-plane patterns are doughnut shaped.

The fabricated antenna and measured gain is shown in Fig. 4.9. A sharp and narrow dip can be observed at notch frequency as antenna does not radiate. The simulated radiation efficiency of the antenna was found to be $\sim$90% throughout the operational band.

To investigate transmission response of the proposed antenna, a pair of antennas are fabricated which are identical in all aspects. The antenna system measurement is carried out outside anechoic chamber considering a real world situation with

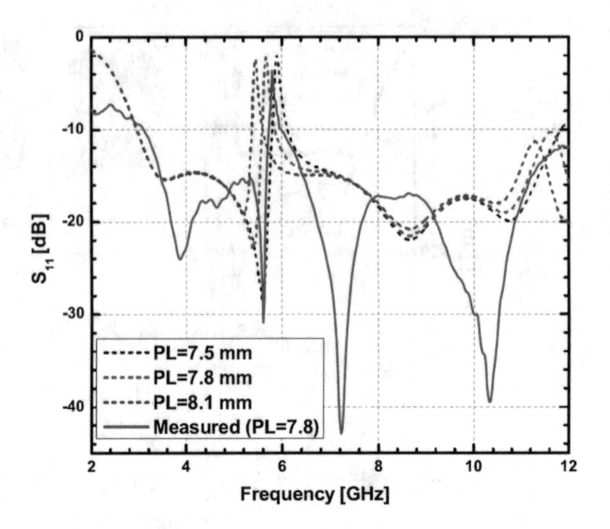

Fig. 4.6 Parametric analysis of ‘PL’and measured return loss [185] [From: J. Malik, Amalendu Patnaik and M.V. Kartikeyan, “Novel band-notch technique for improved time domain performance of printed UWB antenna,” *Proc. of URSI-RCRS 2015, India*. Reproduced courtesy of the URSI-RCRS, Belgium.]

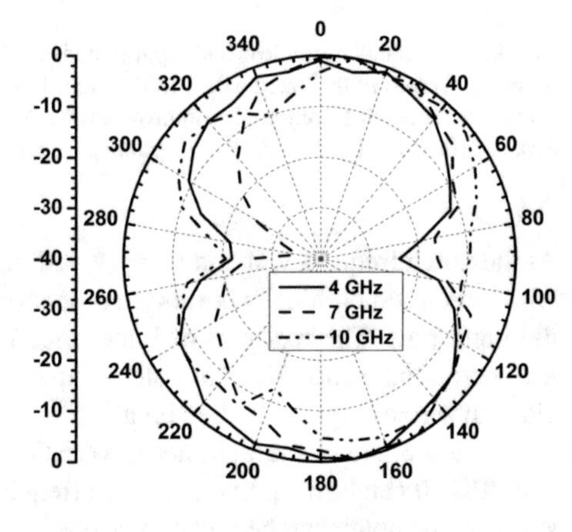

Fig. 4.7 Measured radiation pattern of the antenna in E–plane [185] [From: J. Malik, Amalendu Patnaik and M.V. Kartikeyan, “Novel band-notch technique for improved time domain performance of printed UWB antenna,” *Proc. of URSI-RCRS 2015, India*. Reproduced courtesy of the URSI-RCRS, Belgium.]

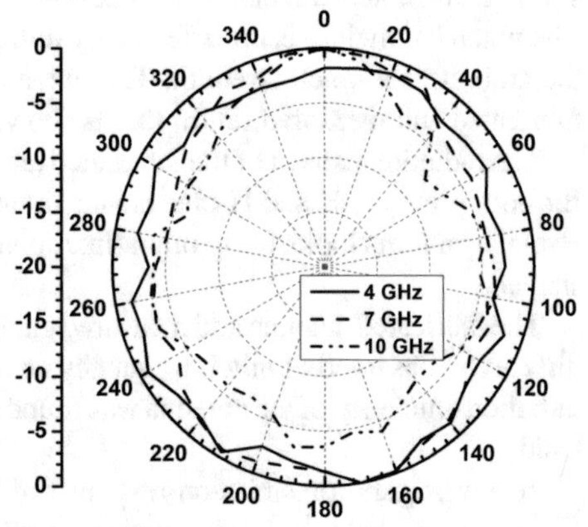

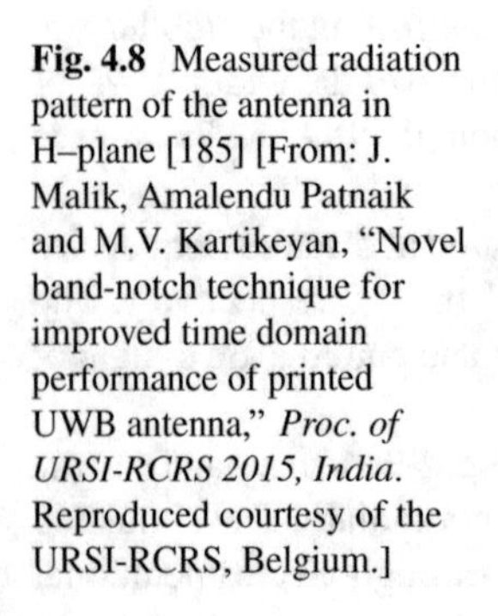
Fig. 4.8 Measured radiation pattern of the antenna in H–plane [185] [From: J. Malik, Amalendu Patnaik and M.V. Kartikeyan, “Novel band-notch technique for improved time domain performance of printed UWB antenna,” *Proc. of URSI-RCRS 2015, India*. Reproduced courtesy of the URSI-RCRS, Belgium.]

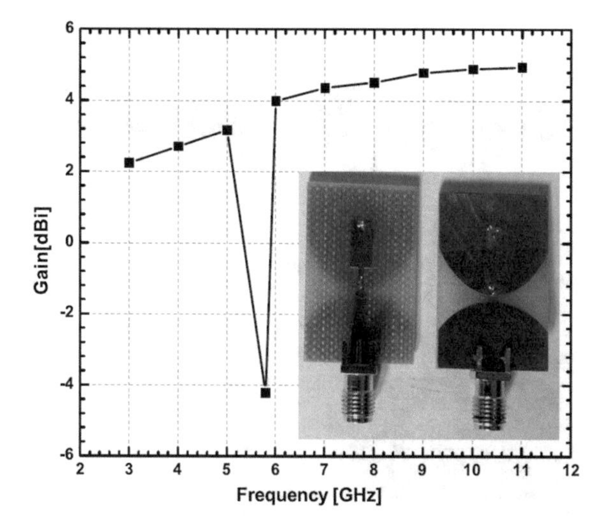

Fig. 4.9 Measured gain and image of fabricated antennas [185] [From: J. Malik, Amalendu Patnaik and M.V. Kartikeyan, "Novel band-notch technique for improved time domain performance of printed UWB antenna," *Proc. of URSI-RCRS 2015, India.* Reproduced courtesy of the URSI-RCRS, Belgium.]

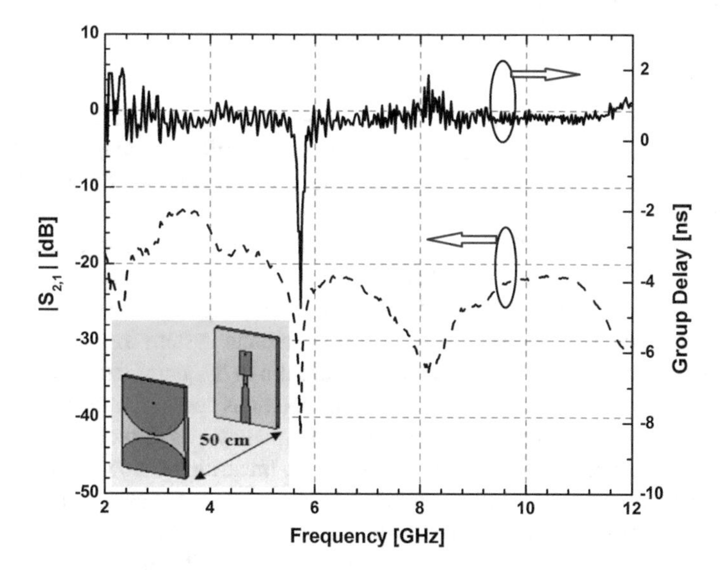

Fig. 4.10 Measured transmission and group delay in face-to-face orientation [185] [From: J. Malik, Amalendu Patnaik and M.V. Kartikeyan, "Novel band-notch technique for improved time domain performance of printed UWB antenna," *Proc. of URSI-RCRS 2015, India.* Reproduced courtesy of the URSI-RCRS, Belgium.]

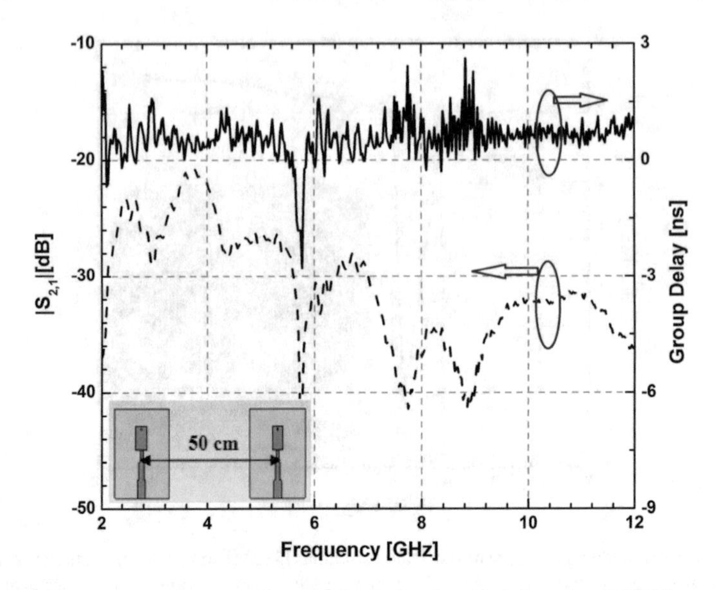

Fig. 4.11 Measured transmission and group delay in side-by-side orientation [185] [From: J. Malik, Amalendu Patnaik and M.V. Kartikeyan, "Novel band-notch technique for improved time domain performance of printed UWB antenna," *Proc. of URSI-RCRS 2015, India*. Reproduced courtesy of the URSI-RCRS, Belgium.]

scatterers and multiple reflecting objects. The entire system consists of two identical band-notched UWB antennas, one as transmitter and other as receiver. The antennas were connected to two ports of VNA using a low loss co-axial cable. Keeping in mind, the limits on maximum output power level of VNA, the antennas were kept apart at a distance of 50 cm in their line of sight (LOS). The measurements were carried out without the use of additional components e.g. power amplifiers at transmitter or low noise amplifier (LNA) at receiver. Figures 4.10 and 4.11 shows measured transmission characteristic of UWB antenna system in face-to-face and side-by-side configuration, respectively. A sharp dip in S_{21} parameter is observed at notch frequency for both the orientations as the antennas does not transmit or receive signal. A sharp change in group delay at notch frequency was also observed for both orientations because of the presence of phase non-linearities at notched-band.

Successful transmission and reception of UWB pulses requires minimum spreading and distortion of the transmitted pulse. This requires that the UWB antennas must be as distortion less as possible. The antennas are tested with an excitation signal as the 4th derivative of the Gaussian pulse. The normalized transmitted and received signal for both orientations of band-notched UWB antenna system are shown in Fig. 4.12. The calculated SFF for the antenna system in face-to-face and side-by-side orientations are 0.947 and 0.938 respectively. The ringing durations in received signal for both cases are significantly small which ensures minimum received pulse distortion.

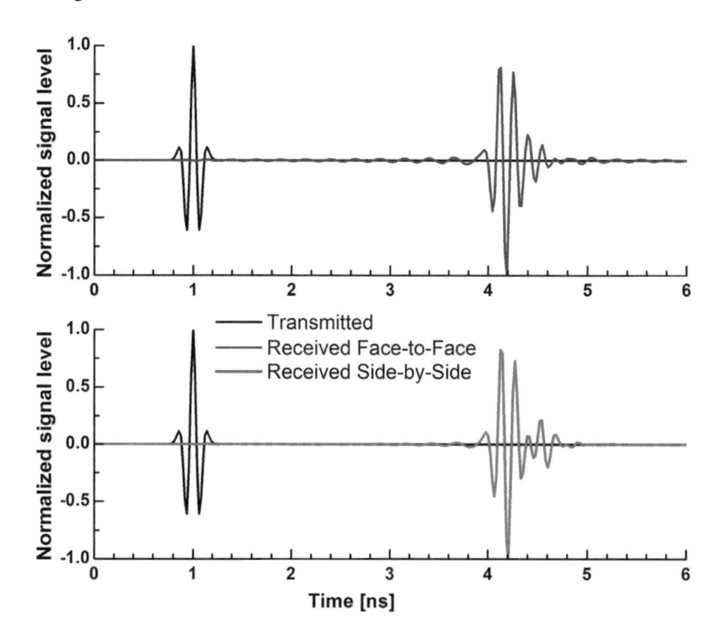

Fig. 4.12 Normalized transmitted and received signals for band-notched UWB antenna system [185] [From: J. Malik, Amalendu Patnaik and M.V. Kartikeyan, "Novel band-notch technique for improved time domain performance of printed UWB antenna," *Proc. of URSI-RCRS 2015, India.* Reproduced courtesy of the URSI-RCRS, Belgium.]

4.3 Concluding Remarks

The present study explores important time-domain behavior of band-notched techniques in UWB antennas for successful application in impulse-radio communications. A comparative analysis of three different band-notch techniques have has been presented. All the techniques have similar kind of frequency domain behavior in terms of rejection bandwidth and strength, but different time-domain performance. Ringing strength and duration for all the 3—techniques have been investigated. To minimize the pulse distortion and ringing duration, a novel band rejection technique using shorted quarter wavelength stub has been discussed. This technique better preserves transmitted pulse shape with minimal distortion and ringing. The notched bandwidth is very narrow and cab be tuned by changing the stub length. Measured results show gain flatness and well behaved omnidirectional radiation patterns in H-plane over the entire operational band. The time domain performance parameters e.g. group delay, SFF and ringing duration for proposed antenna system in both orientations have been evaluated and discussed.

Chapter 5
Printed Antennas for MIMO: Exploitation of Pattern Diversity

5.1 Design and Analysis of MIMO Antenna with Omnidirectional Pattern Diversity

5.1.1 Introduction and Related Work

MIMO is the key technology that enables implementation of future higher data rate communications without increasing either spectral bandwidth or transmit power level. It takes the advantage of multiple antennas at transmitter and/or receiver to increase channel capacity. Using multiple antennas, parallel independent wireless channels can be realized in a rich-multipath environment for transmitting and receiving data streams that yields a linear increase in the channel capacity. The idea is to create parallel resolvable channels for receiving uncorrelated signals at the receiver. To receive uncorrelated signals, MIMO with spatial diversity is a good solution. However, in a mobile terminal space is a constraint, and antenna elements can't be placed far apart to enjoy spatial diversity. So to create parallel channels, MIMO with pattern and/or polarization diversity is the best option. Successful implementation of MIMO requires low correlation between these channels i.e. high isolation between the antenna elements in the system is a must to achieve the desired diversity performance. Popular practices like use of neutralization line [97], application of metamaterials [96], use of parasitic resonators [99, 101], and using ground slot technique [100] are proved for achieving high isolation. A close observation reveals that these techniques require extra space to be implemented. Again these techniques use resonant type structures with frequency same as that of the radiating antenna elements. This reduces the radiation and overall efficiency of the system. Other techniques like exciting orthogonal modes simultaneously at overlap frequency [106, 107] have been reported. However, these designs have disadvantages as these are bulky in nature, with different impedance matching/bandwidth and radiation characteristics for different ports/elements. Here a planar printed dual-element MIMO antenna is proposed with omnidirectional radiation patterns in orthogonal planes.

J. Malik et al., *Compact Antennas for High Data Rate Communication*,
Springer Topics in Signal Processing 14, DOI 10.1007/978-3-319-63175-2_5

Table 5.1 Design parameters and values of MIMO antenna with pattern diversity [186] [From: J. Malik, D. Nagpal and M.V. Kartikeyan, "MIMO Antenna with Omnidirectional Pattern Diversity," *Electronics Letters*, vol. 52, No. 2, pp. 102–104, 2016. © IET 2016. Reproduced courtesy of the IET, UK]

Parameter	G1	G2	W1	L1	L2
Value (mm)	0.4	0.2	7.4	11.0	10.0

Table 5.2 Resonance behaviour of antenna with capacitance [186] [From: J. Malik, D. Nagpal and M.V. Kartikeyan, "MIMO Antenna with Omnidirectional Pattern Diversity," *Electronics Letters*, vol. 52, No. 2, pp. 102–104, 2016. © IET 2016. Reproduced courtesy of the IET, UK]

Capacitance (pF)	0.1	0.2	0.6	0.8	1.0	1.4	2.0	2.4
Frequency (GHz)	3.31	3.0	2.68	2.6	2.53	2.46	2.29	2.14

5.1.2 Antenna Design and Implementation

Figure 5.1 shows the schematic view of the proposed MIMO antenna system. The MIMO system consists of two dipole-like radiators placed orthogonal to each other and printed on the opposite sides of a dielectric substrate. Low cost FR-4 is used as the substrate material with a thickness of 1.524 mm and dielectric constant as 4.4 ($\tan\delta = 0.002$). The overall footprint of the system is $(45 \times 45 \times 1.524)\,\text{mm}^3$. The MIMO antenna system starts with a simple printed dipole of length L1 = 11.0 mm and width W1 = 2.2 mm. A gap G1 = 0.4 mm separates the two arms of the dipole. The design parameters are given in Table 5.1. Upon exciting the dipole with a 50 Ω discrete port, the resonance frequency comes to be 4.38 GHz. It is shown in Fig. 5.2 (inset 'A'). In the second step, a square patch like element is added to both ends (Fig. 5.2, inset 'B') of the dipole with a gap of G2 = 0.2 mm. This shifts the resonance at 3.5 GHz. To further reduce the resonance frequency, RF capacitors is used between the dipole arms and the loaded square element as shown in Fig. 5.1. The resonance frequency of the antenna can be tuned by changing the capacitance of the capacitor. Table 5.2 shows the variation of resonance frequency with capacitance. It can be observed that the peak of resonant frequency varies from 3.31 to 2.14 GHz by tuning the capacitance from 0.1 to 2.5 pF. To realize a 2-element MIMO antenna with pattern diversity, a second antenna similar to the first antenna is placed on the back side of the substrate with a 90^0 orientation between them (Fig. 5.1). As shown in Fig. 5.1, the top dipole is oriented along X-axis where as the bottom dipole is oriented along Y-axis. Since the dipoles are similar in all manners, the resonance behavior S11 and S22 are same. Due to the orthogonal orientation, the coupling (S12/S21) between the antenna elements is very low (Fig. 5.2).

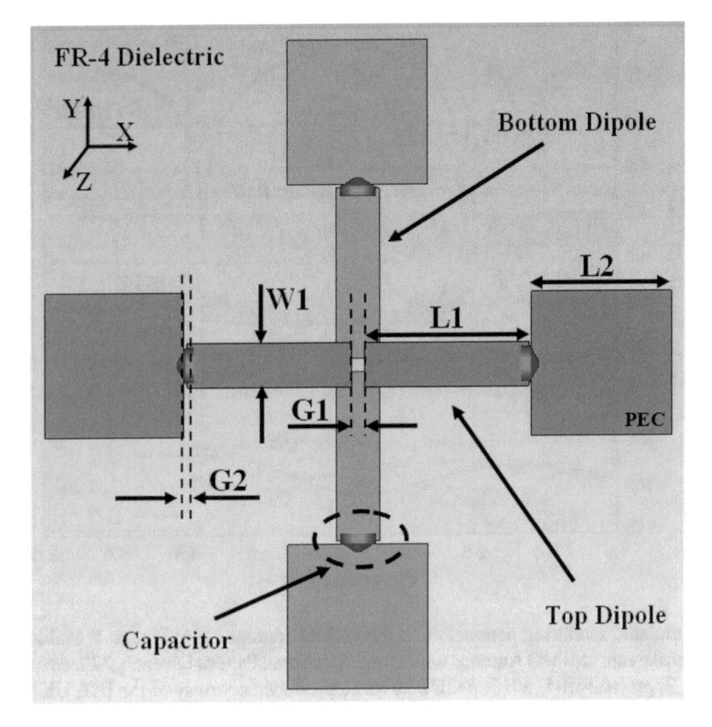

Fig. 5.1 Schematic view of the proposed MIMO antenna [186] [From: J. Malik, D. Nagpal and M.V. Kartikeyan, "MIMO Antenna with Omnidirectional Pattern Diversity," *Electronics Letters*, vol. 52, No. 2, pp. 102–104, 2016. © IET 2016. Reproduced courtesy of the IET, UK]

5.1.3 Results and Discussion

The S-parameters are measured with an Agilent PNA network analyzer. Figure 5.3 shows the measured scattering parameters of the fabricated antenna with a 1.2 pF loaded capacitor. The photograph of the fabricated antenna is shown in the inset. It can be observed that the measured −10 dB impedance bandwidth for both antenna elements covers the lower WLAN (2.4–2.484 GHz) band with good impedance matching. The measured coupling S21/S12 is well below −50 dB.

Figure 5.4 shows the simulated 3D radiation pattern of both antennas. The pattern diversity can be clearly observed from the figures. The antennas have dough-nut shape pattern orthogonal to each other. When upper antenna is excited (port-1), the radiation pattern is omnidirectional in the YZ-plane with nulls along X-axis. Whereas, the radiation pattern for bottom antenna (port-2) is omnidirectional in XZ-plane with nulls along Y-axis. The measured radiation pattern at 2.4 GHz in E-plane and H-plane are shown in Figs. 5.5 and 5.6 respectively. The measured radiation patterns are in good agreement with the simulated patterns. The gain measurement is carried out in anechoic chamber against a standard gain horn antenna using substitution/transfer

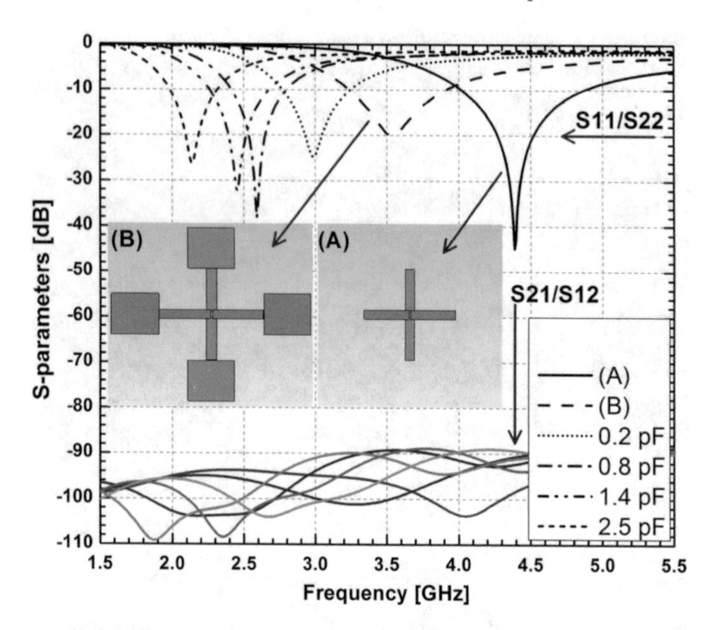

Fig. 5.2 Simulated scattering parameters of the MIMO antenna [186] [From: J. Malik, D. Nagpal and M.V. Kartikeyan, "MIMO Antenna with Omnidirectional Pattern Diversity," *Electronics Letters*, vol. 52, No. 2, pp. 102–104, 2016. © IET 2016. Reproduced courtesy of the IET, UK]

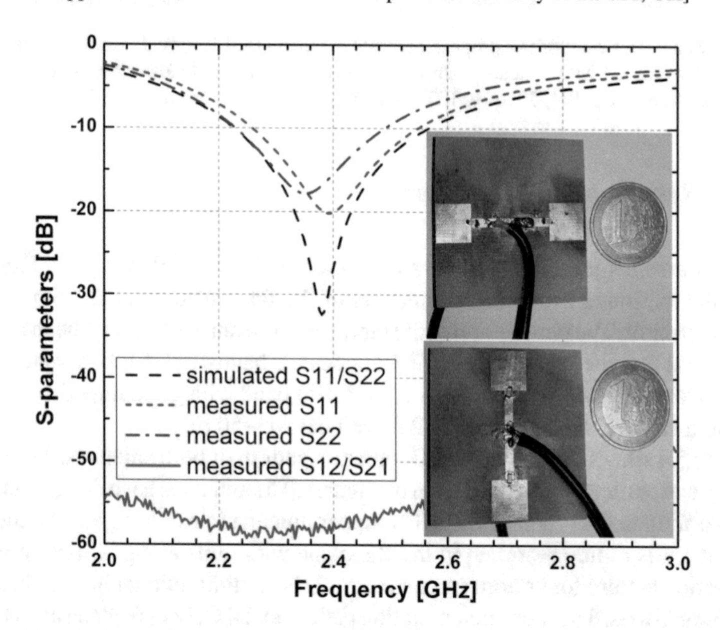

Fig. 5.3 Measured scattering parameters of the MIMO antenna with photograph of the fabricated antenna [186] [From: J. Malik, D. Nagpal and M.V. Kartikeyan, "MIMO Antenna with Omnidirectional Pattern Diversity," *Electronics Letters*, vol. 52, No. 2, pp. 102–104, 2016. © IET 2016. Reproduced courtesy of the IET, UK]

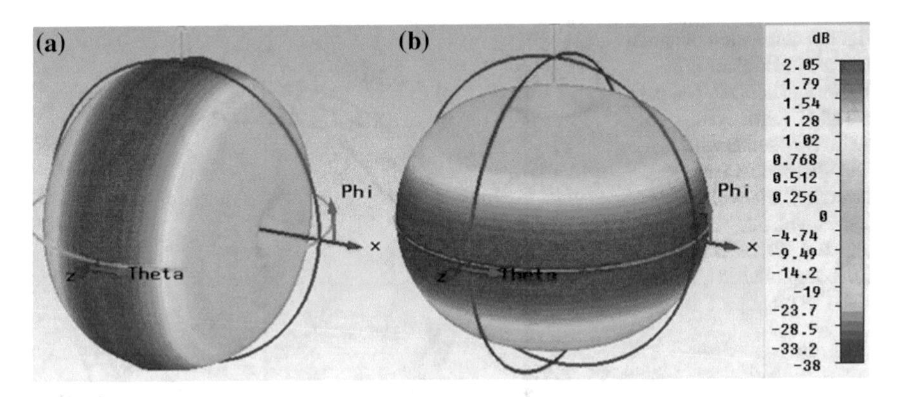

Fig. 5.4 Simulated 3D radiation pattern for **a** port-1: *top* antenna **b** port-2: *bottom* antenna [186] [From: J. Malik, D. Nagpal and M.V. Kartikeyan, "MIMO Antenna with Omnidirectional Pattern Diversity," *Electronics Letters*, vol. 52, No. 2, pp. 102–104, 2016. © IET 2016. Reproduced courtesy of the IET, UK]

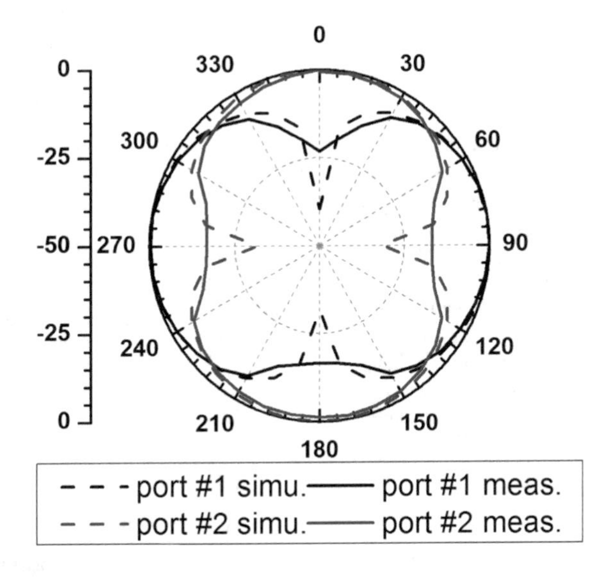

Fig. 5.5 Measured radiation patterns in E–plane [186] [From: J. Malik, D. Nagpal and M.V. Kartikeyan, "MIMO Antenna with Omnidirectional Pattern Diversity," *Electronics Letters*, vol. 52, No. 2, pp. 102–104, 2016. © IET 2016. Reproduced courtesy of the IET, UK]

method. The measured gain for both the antenna is similar and shown in Fig. 5.7. The ECC is the measure to evaluate the diversity performance of an MIMO antenna system. The correct estimate of ECC is ideally calculated from 3D radiation pattern. Alternately it can be calculated from the scattering parameters assuming the MIMO system will operate in a uniform multipath rich environment. In simulation the ECC level is close to zero due to very high isolation between the elements/ports with orthogonal pattern orientation. The measured ECC level is 0.0043 in the target 2.4 GHz band which ensures good diversity operation.

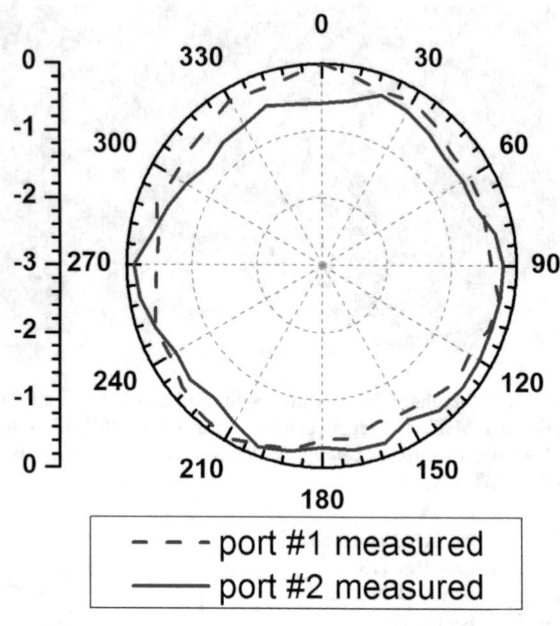

Fig. 5.6 Measured radiation patterns in H–plane [186] [From: J. Malik, D. Nagpal and M.V. Kartikeyan, "MIMO Antenna with Omnidirectional Pattern Diversity," *Electronics Letters*, vol. 52, No. 2, pp. 102–104, 2016. © IET 2016. Reproduced courtesy of the IET, UK]

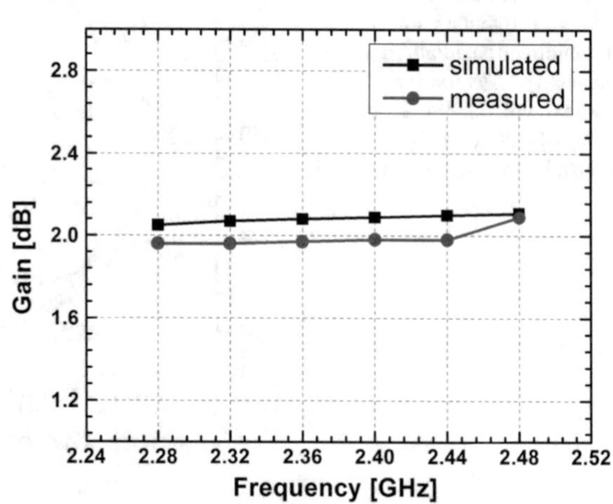

Fig. 5.7 Simulated and measured gain of proposed MIMO antenna [186] [From: J. Malik, D. Nagpal and M.V. Kartikeyan, "MIMO Antenna with Omnidirectional Pattern Diversity," *Electronics Letters*, vol. 52, No. 2, pp. 102–104, 2016. © IET 2016. Reproduced courtesy of the IET, UK]

5.2 Design and Analysis of MIMO Antenna with Pattern and Polarization Diversity

5.2.1 Introduction and Related Work

Antenna mutual coupling is a key issue of concern and a primary limiting factor for successful MIMO operation [12]. Integration of two antennas operating at same frequency with low mutual coupling in a MIMO system is quite a challenge under compactness constraint. For the same, isolation enhancement techniques e.g. using decoupling network [97, 98], using neutralization line joining antenna elements [174], using parasitic coupling elements between antenna elements [99, 101] and using ground slot technique [100] are popular in practice. Further compact MIMO system can be achieved by eliminating extra circuitry needed to reduce the correlation, using pattern and/or polarization diversity. These utilize orthogonal patterns and orthogonal polarizations as a means to create uncorrelated channels [102–107]. In [102], a mono-cone antenna loaded with a cup-shaped patch and shorting metallic posts for broadband MIMO operation with pattern diversity is presented. In [104], authors demonstrated a technique for polarization diversity by placing three linearly polarized printed dipole antennas physically orthogonal to each other. Use of orthogonal feeding to a planar inverted F-antenna (PIFA) [103], and use of modified shorted bow-tie antenna with two different feeding mechanism [105] are presented for combined pattern and polarization diversity. Exciting TM_{01} and TM_{11} modes simultaneously at overlap frequency in a hybrid-fed circular patch [106] and in a short-circuited ring patch [107] results excellent port-to-port isolation with combined pattern and polarization diversity.

On keen observation of the MIMO antenna structures reported in above referenced literature, the distance between individual radiators and use of a separate isolation technique are key factors that decide size of overall MIMO antenna system. The present work deals with the design of a compact MIMO antenna with decoupled radiators utilizing simultaneous pattern and polarization diversity. One of the radiators is designed for circular polarization and other one is for linear polarization. The designed MIMO antenna also possesses good pattern diversity due to the orthogonal orientations between radiation maxima of the radiators. The detail design methodology is presented, and the simulation and measurement results are discussed to validate the diversity performance of proposed MIMO antenna.

5.2.2 Antenna Design and Implementation

Design parameters of the proposed MIMO antenna is shown in Fig. 5.8. The resonators are printed on the opposite sides of FR-4 substrate ($\varepsilon_r = 4.4$ and thickness of 1.524 mm). The dimensions of all physical parameters of the proposed antenna are given in Table 5.3. The upper radiator is a square patch element with edge fed, through a quarter wave line. The dimensions of the square patch are calculated using transmission line theory for given substrate specifications. The square patch is incorporated with corner chamfer and 45^0 slant slot as shown in Fig. 5.8a to get circular polarization. An off-center feed is used to get good axial ratio. Here it may be noted that, moving feed-line about the center of patch along the edge, impedance bandwidth is not much affected but axial ratio and 3 dB axial ratio bandwidth are

Table 5.3 Design parameters and values of MIMO antenna with combined diversity [187] [J. Malik, A. Patnaik and M.V. Kartikeyan, "Novel Printed MIMO Antenna with Pattern and Polarization Diversity," *IEEE Antennas and Wireless Propagation Letters*, vol. 14, pp. 739-742, 2015. © IEEE 2015. Reproduced courtesy of the IEEE, USA]

Parameter	PL	PW	L	P	R	FL1	FL2	FD1	FD2	SW
Value (mm)	11.6	11.6	6.8	0.5	3.0	3.1	6.1	3.0	0.9	30
Parameter	SL	X1	X2	Y1	Y2	Y3	L1	W	G	
Value (mm)	25	0.5	6.0	6.0	6.0	2.2	12.3	1.0	0.4	

greatly affected. The final optimized position (2.4 mm from the center-line of patch) of feed-line is selected by parametric analysis. Figure 5.8b shows bottom radiator of the MIMO antenna system. The bottom radiator is a modified inter–digital type structure and fed by a coplanar slot line. The length of each finger is half a wavelength corresponding to WLAN (5.8 GHz) frequency. The points marked and are grounding and signal points respectively for the discrete port used in simulation. The length of the tapered section is optimized parametrically to achieve optimum impedance matching in desired band.

5.2.3 Results and Discussion

CST Microwave studio v12 is used to design, simulate and analyze the proposed MIMO antenna. Simulated S-parameters are shown in Fig. 5.9. Isolation of the order of -13 dB between two radiators over desired operational band can be marked from the figure. Here, it may be noted that we have not used any isolation enhancement technique as such. The proposed structure is fabricated and the S-parameters are measured using Agilent PNA series network analyzer to cross verify the simulated results. Figure 5.10 shows the measured scattering parameters. The resonant frequencies of both radiators are observed to have little right shift towards higher frequency, still the measured impedance bandwidth covers the desired WLAN band. Little improvement in measured port-to-port isolation is observed compared to simulation.

A parametric analysis is carried out to investigate the circular polarization (CP) performance (axial ratio and CP bandwidth) of the antenna along the broadside direction. Figure 5.11a shows the effect of varied 'R'on axial ratio and optimum value of 3.0 mm is chosen. Figure 5.11b shows the parametric analysis of 'L'on CP performance. The optimum value is chosen as 6.8 mm. The axial ratio is measured at three different frequencies inside anechoic chamber. The measured axial ratio at 5.8, 5.9 and 6.0 GHz are 1.92, 2.23 and 2.61 respectively.

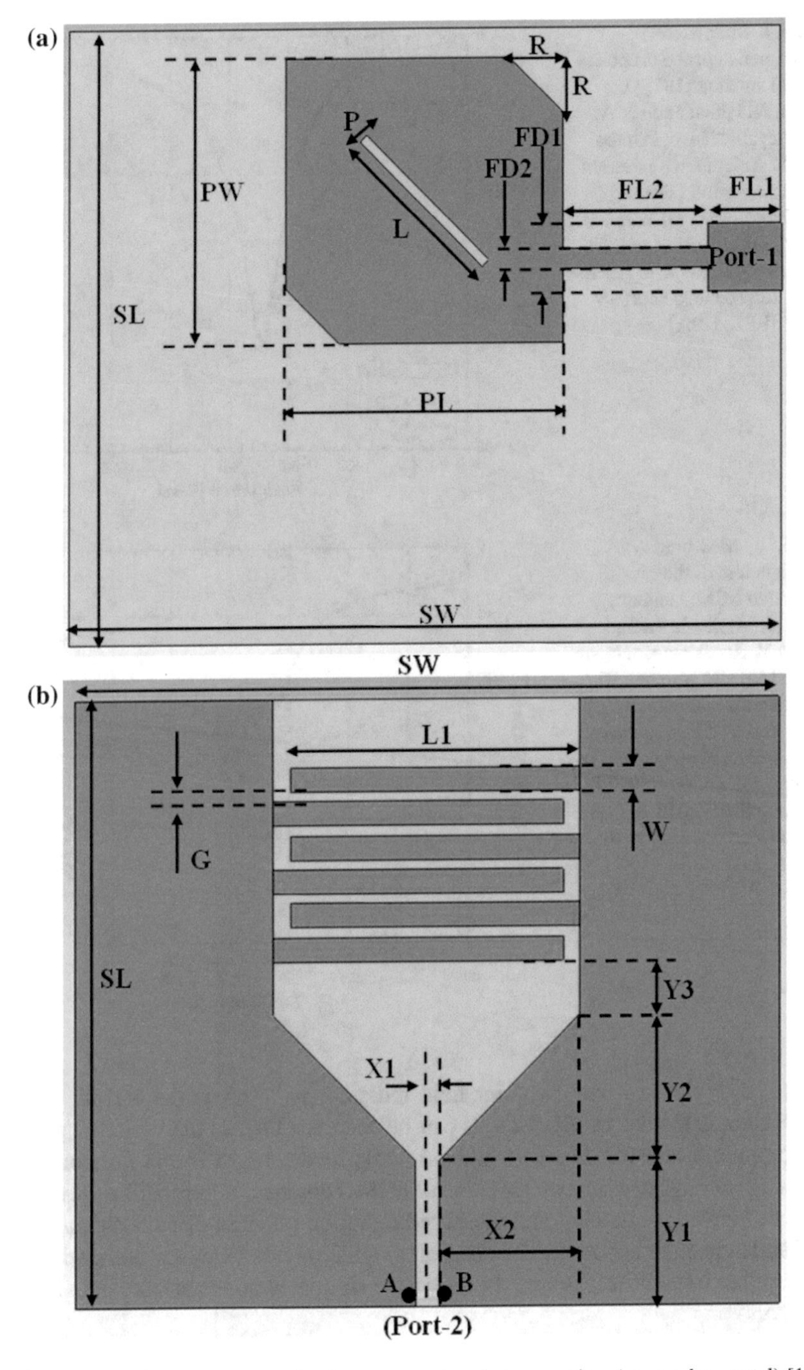

Fig. 5.8 Schematic of the proposed antenna **a** *top* view **b** *bottom* view (*grey color*: metal) [187] [J. Malik, A. Patnaik and M.V. Kartikeyan, "Novel Printed MIMO Antenna with Pattern and Polarization Diversity," *IEEE Antennas and Wireless Propagation Letters*, vol. 14, pp. 739-742, 2015. © IEEE 2015. Reproduced courtesy of the IEEE, USA]

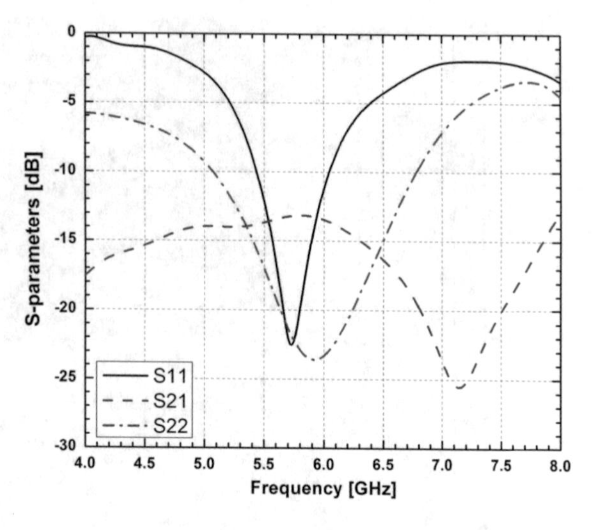

Fig. 5.9 Simulated S-parameters of the proposed MIMO antenna [187] [J. Malik, A. Patnaik and M.V. Kartikeyan, "Novel Printed MIMO Antenna with Pattern and Polarization Diversity," *IEEE Antennas and Wireless Propagation Letters*, vol. 14, pp. 739–742, 2015. © IEEE 2015. Reproduced courtesy of the IEEE, USA]

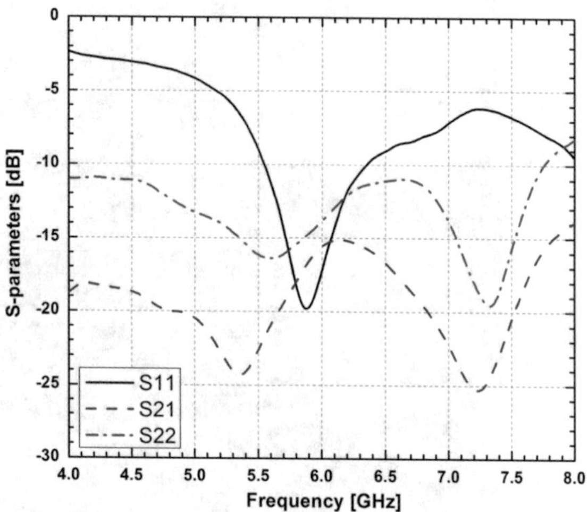

Fig. 5.10 Measured S-parameters of the fabricated MIMO antenna [187] [J. Malik, A. Patnaik and M.V. Kartikeyan, "Novel Printed MIMO Antenna with Pattern and Polarization Diversity," *IEEE Antennas and Wireless Propagation Letters*, vol. 14, pp. 739–742, 2015. © IEEE 2015. Reproduced courtesy of the IEEE, USA]

The 3-D view of simulated far field radiation patterns of the MIMO antenna at 5.8 GHz is shown in Fig. 5.12. It can be seen for Port-1, the radiation maxima is along the broadside direction (z-axis) while for Port-2, it is end fire confirming pattern diversity between two radiators of MIMO antenna. Figure 5.13 shows a comparison between simulated and measured radiation patterns of both radiators. The agreement is quite apparent. The simulated radiation efficiency for both radiators is observed to be $\sim 96\%$. Figure 5.14 shows the simulated and measured gain for both radiators.

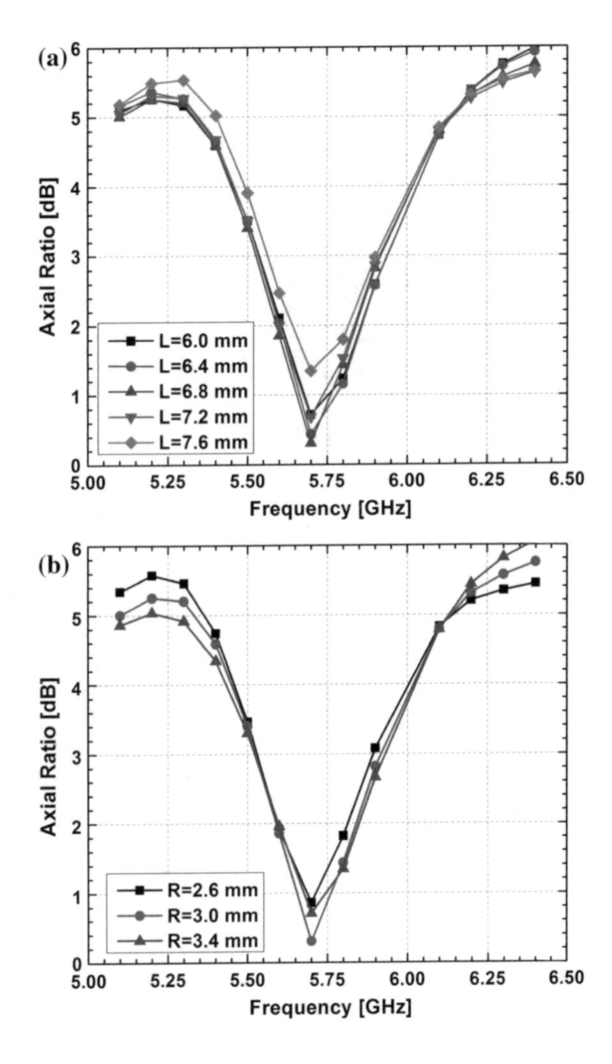

Fig. 5.11 Simulated parametric analysis of axial ratio for port-1 with variation in **a** chamfer length 'R' **b** slot length 'L'[187] [J. Malik, A. Patnaik and M.V. Kartikeyan, "Novel Printed MIMO Antenna with Pattern and Polarization Diversity," *IEEE Antennas and Wireless Propagation Letters*, vol. 14, pp. 739–742, 2015. © IEEE 2015. Reproduced courtesy of the IEEE, USA]

The envelope correlation coefficient (ECC) is a measure to evaluate the diversity performance of a MIMO antenna. The correct estimate of ECC is ideally calculated from 3-D radiation pattern. Assuming that the MIMO antenna will operate in a uniform multipath rich environment, it can be alternatively calculated by using the scattering parameters. The computed ECC for the proposed MIMO system is well below 0.06 for the desired band, which ensures a good diversity performance.

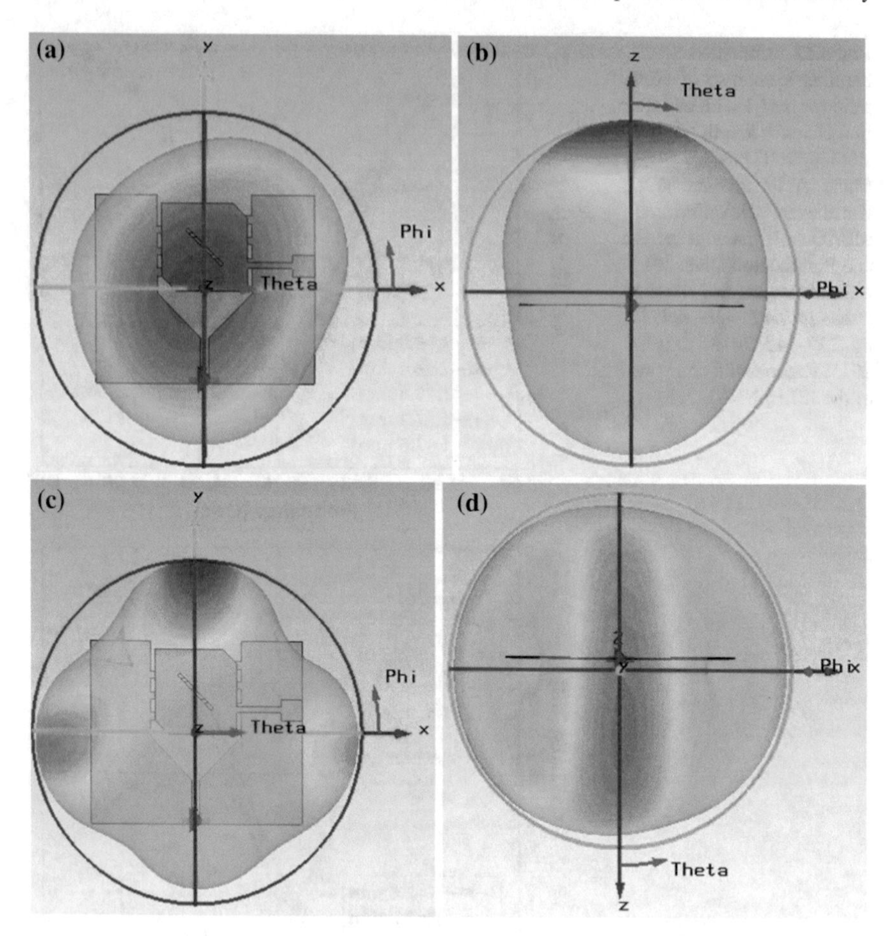

Fig. 5.12 Simulated 3-D far field radiation pattern at 5.8 GHz for (**a**, **b**) port-1 excited (**c**, **d**) port-2 excited [187] [J. Malik, A. Patnaik and M.V. Kartikeyan, "Novel Printed MIMO Antenna with Pattern and Polarization Diversity," *IEEE Antennas and Wireless Propagation Letters*, vol. 14, pp. 739–742, 2015. © IEEE 2015. Reproduced courtesy of the IEEE, USA]

5.3 Concluding Remarks

In this chapter, compact MIMO antennas with pattern and polarization diversity have been presented and discussed. In the first design, a dual-port planar MIMO antenna designed for lower WLAN band (2.4 GHz) applications is presented. In measurement both the antennas show a wide impedance bandwidth with in-band isolation better than 50 dB between them. Orthogonally oriented radiation patterns are suitable for a pattern diversity MIMO application. With an external loaded capacitor, the resonance frequency of the MIMO antenna can be tuned over a wide range of frequency. In the second design a compact MIMO antenna with simultaneous pattern and polarization

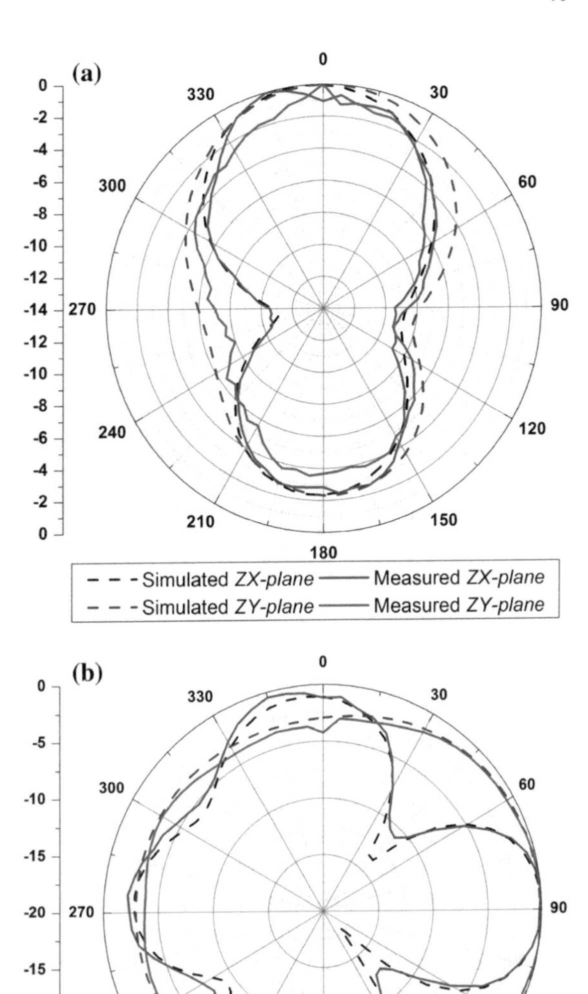

Fig. 5.13 Simulated and measured radiation pattern at 5.8 GHz for **a** Port-1 **b** port-2 [187] [J. Malik, A. Patnaik and M.V. Kartikeyan, "Novel Printed MIMO Antenna with Pattern and Polarization Diversity," *IEEE Antennas and Wireless Propagation Letters*, vol. 14, pp. 739–742, 2015. © IEEE 2015. Reproduced courtesy of the IEEE, USA]

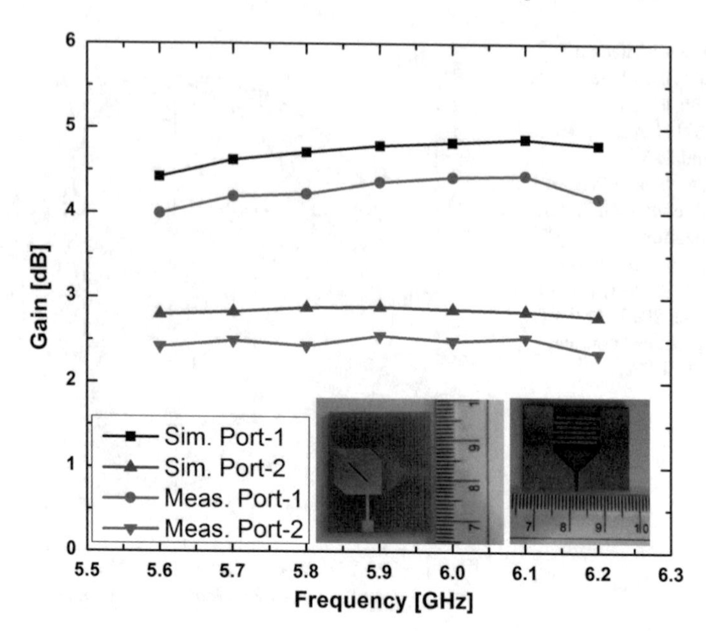

Fig. 5.14 Fabricated antenna and measured gain for both resonators [187] [J. Malik, A. Patnaik and M.V. Kartikeyan, "Novel Printed MIMO Antenna with Pattern and Polarization Diversity," *IEEE Antennas and Wireless Propagation Letters*, vol. 14, pp. 739–742, 2015. © IEEE 2015. Reproduced courtesy of the IEEE, USA]

diversity has been proposed for WLAN applications. The MIMO antenna system is compact in the sense that the two radiators are physically separated by a distance equal to the thickness of the substrate material used. Though, the antennas are placed very close to each other, the coupling between them is low. The upper resonator has been designed to have right-handed circular polarization performance, while the bottom radiator as linearly polarized. The far field radiation pattern for both radiators is orthogonal to each other, ensuring good pattern diversity performance. Design methodology for the proposed MIMO antenna system is simple and can be extended to design antenna for other frequencies with suitable scaling, parametric analysis and optimization.

Chapter 6
Printed Antennas for MIMO: Exploitation of Polarization Diversity

6.1 Introduction and Related Work

In indoor environment, multiple copies of transmitted signal arrive at the receiver with different time delay due to reflection and diffraction by the surrounding objects, walls and ceiling. The ensemble of these signals produces fading over time that varies with spatiotemporal scenario of the channel. Antenna diversity is a communication method that takes the advantage of multiple antennas at transmitter and/or receiver to capture statistically independent copies of transmitted signal. At the receiver end, various signal combining methods results in a diversity gain that increases the radio link quality. In [114], authors presented the probability of depolarization of the transmitted signal due to multiple reflections from sharp corners and multi-layered walls in indoor environment. Use of polarization diversity concept in high multipath and depolarized indoor environment shows excellent immunity to fading [115]. Circularly polarized antennas are widely used to solve the problems of polarization mismatch between received and transmitted signal. Furthermore, dual circularly polarized antennas are generally utilized in MIMO system to avoid fading loss, frequency reuse and increase system capacity. Antennas with circular polarization diversity have been widely studied, in which patch antenna is the most popularly investigated for its low-profile and ease of fabrication.

In [116–119], authors have reported excellent MIMO antenna topologies for circular polarization diversity applications. A close investigation from fabrication angle reveals bulky nature [116], complicated feeding network [117], complexity due to active components [118] and large separation between the radiating elements [119], in these reported antenna limits the practical applicability. The present design is planar and simple without any complicated feeding structure. It is also compact in the sense that it uses only one radiating patch element for MIMO configuration.

Here, we discuss a thoughtful technique to generate circular polarization in a square patch using L-type feeding line. The design concept is extended to realize a MIMO antenna having circular polarization diversity with compact structure. A decoupling network is incorporated to enhance port-to-port isolation for good

J. Malik et al., *Compact Antennas for High Data Rate Communication*,
Springer Topics in Signal Processing 14, DOI 10.1007/978-3-319-63175-2_6

Table 6.1 Design parameters and values of MIMO antenna with CP diversity [188] [J. Malik, A. Patnaik and M.V. Kartikeyan, "Novel Printed MIMO Antenna with Dual-Sense Circular Polarization Diversity", *Proc. of URSI-RCRS 2015, India*. Reproduced courtesy of the URSI-RCRS, Belgium]

Parameter	SW	SL	R	G	L1	L2	L3
Value (mm)	25	25	2.8	0.2	16.4	0.6	8.8
Parameter	L4	L5	L6	L7	L8	L9	L10
Value (mm)	4.8	6.6	2.6	3.0	2.2	6	10.9

diversity gain. The detail antenna design, simulation and measurement results are described in the subsequent sections. The measured results are in agreement with the expected data obtained from the full wave simulation.

6.2 Antenna Design and Implementation

The geometrical configuration of the proposed MIMO antenna system is shown in Fig. 6.1. The values of various physical design parameters of the antenna are given in Table 6.1. The MIMO system consists of a corner-truncated square patch antenna with a center shorting pin with ground, two L-shaped feed-lines placed symmetrically to the patch, and a decoupling network between the two excitation ports. The conceptual design of the MIMO antenna started with an aim to excite single sense of circular polarization in a square patch antenna using a L-type feed-line as shown in Fig. 6.1. The initial dimension of the patch was calculated using transmission line model [120]. The aim of the present design is to excite both sense of circular polarization in single patch. The gap-coupled feeding mechanism is preferred in this case over other feeding techniques like coaxial, or microstrip feeding. Since in later cases, due to direct physical contact of both SMA connectors to same patch element and ground, it would be rather difficult to control the surface current coupling between the feeding/excitation ports. A uniform gap of 0.15 mm is maintained between patch and feed-line.

For each port, the two branches of the L-type feed-line induces two orthogonal components of the electric field on the patch as shown in Fig. 6.1. When port-1 is excited, $E1_X$ and $E1_Y$ are the two components of induced field. Similarly when port-2 is excited, $E2_X$ and $E2_Y$ are the two orthogonal components of the induced field on the patch. A phase quadrature can be achieved between two components of induced field by tuning the relative lengths of two arms 'L5' and 'L6', which is essential to generate the intended circular polarization [121]. Parametric analysis is done to tune the lengths of the orthogonal arms to produce circular polarization, which is confirmed by observing the simulated axial ratio. To further enhance the CP behavior, the corners of the square patch are chamfered with a length 'R'. A shorting pin is

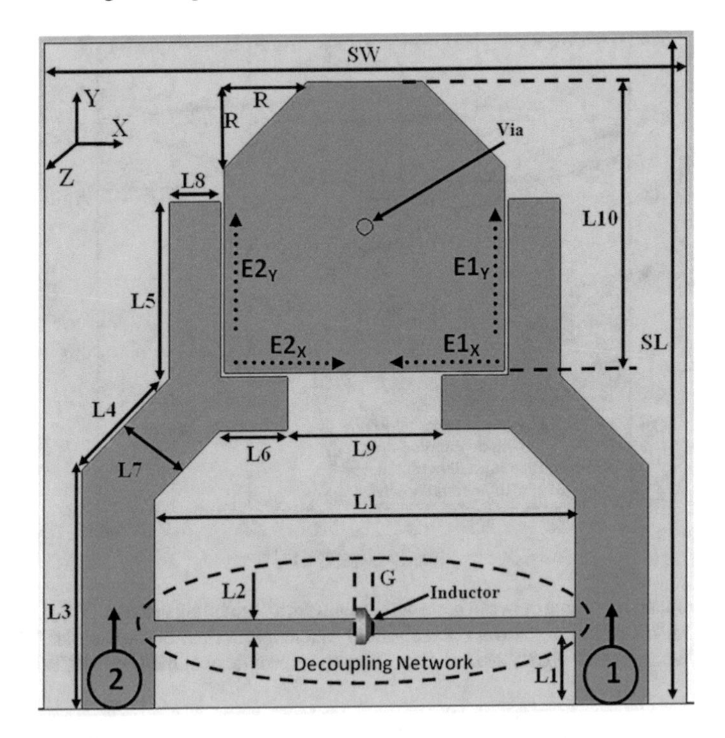

Fig. 6.1 Schematic diagram of proposed MIMO antenna (*dotted arrows* show the induced fields) [188] [J. Malik, A. Patnaik and M.V. Kartikeyan, "Novel Printed MIMO Antenna with Dual-Sense Circular Polarization Diversity", *Proc. of URSI-RCRS 2015, India*. Reproduced courtesy of the URSI-RCRS, Belgium]

placed at the center of the patch and connected to the ground. It suppresses excitation of undesirable modes that improves the axial ratio and minimize cross-polarization radiation in the broadside direction.

In the present MIMO antenna system, single patch radiator supports both sense of circular polarization simultaneously. The coupling between the two excitation ports is controlled by two physical mechanisms, i.e. coupling of radiated field components and coupling due to induced surface current components. Ideally there should not be any coupling between radiated LHCP and RHCP fields, due to their orthogonal nature. So the surface current coupling is dominant. The X-axis component of the two induced fields from the feed-lines are out of phase ($\Delta\phi = 180^0$), whereas the Y-axis components of the induced fields are in-phase ($\Delta\phi = 0^0$). Therefore the ports will be tightly coupled. To enhance the port-to-port isolation, a decoupling network is incorporated to the original design. In literature many decoupling networks have been proposed to enhance the port-to-port isolation. Ground slits are proved to disperse the surface current coupling effectively at the desired frequency band [97, 122, 123]. But this is not a viable solution for the present case, as protruding slots in the ground plane or underneath the patch will eventually disturb the CP behavior significantly. In

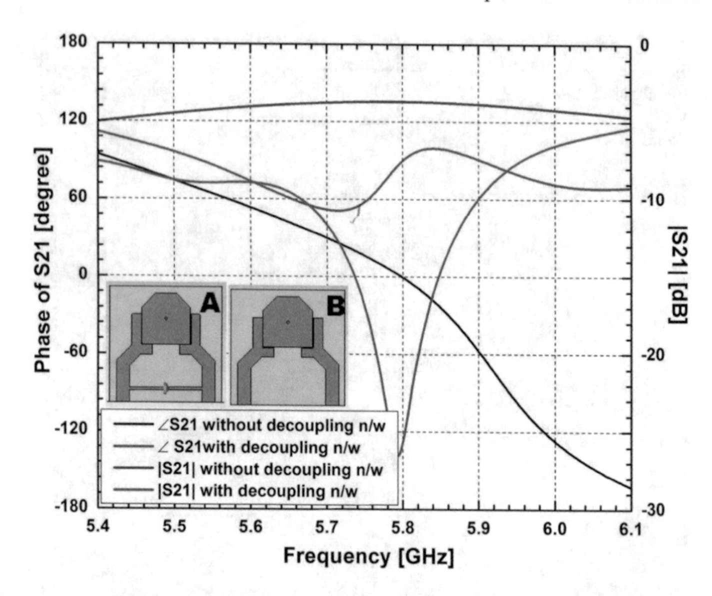

Fig. 6.2 Simulated coupling between ports without and with decoupling network [188] [J. Malik, A. Patnaik and M.V. Kartikeyan, "Novel Printed MIMO Antenna with Dual-Sense Circular Polarization Diversity", *Proc. of URSI-RCRS 2015, India*. Reproduced courtesy of the URSI-RCRS, Belgium]

[97] authors presented an excellent idea to design decoupling network for strongly coupled antennas, but at the cost of complexity. Method like use of metamaterial for isolation enhancement has been presented in [124], but useful only for specific designs.

The coupling effect depends on the excitation phase to the elements in a multi-antenna system. Usually in-phase coupling increases the correlation between the elements. Isolation can be improved by using transmission lines of quarter wavelength between the feeding lines of the coupled antennas [125]. This has a limitation of large size decoupling network. The decoupling network can be further miniaturized using lumped elements. In the present case the idea is to use a decoupling network at specific distance from the antennas input ports, where the equivalent mutual impedance is purely imaginary. The decoupling network consists of a neutralization line with gap 'G'and loaded with an inductor of 1.2 nH. This improves the isolation of the order of -25 dB in simulation. The simulated phase and magnitude of coupling coefficient for the present MIMO antenna with and without decoupling network is shown in Fig. 6.2. Without the decoupling network, the ports are tightly coupled as the phase angle is 0^0. Introducing the decoupling network, phase quadrature is achieved between the two ports that lowered the coupling. The location of the line from port (L1) and value of the lumped inductor are optimized parametrically to achieve phase quadrature.

6.3 Simulation and Measurement Results

The full-wave simulation of the proposed MIMO antenna is carried out using the transient solver in CST Microwave Studio v12. The simulated antenna after detail parametric analysis, tuning and optimization is fabricated on a low cost FR-4 substrate ($\varepsilon_r = 4.4$, tan$\delta = 0.0024$, h $= 1.524$ mm). The simulated and measured scattering parameters of the antenna are shown in Fig. 6.3. Due to symmetric nature of the MIMO antenna, S11 and S21 are shown.

Figure 6.4 shows the simulated and measured axial ratio. The axial ratios are measured inside anechoic chamber. The antenna is targeted for IEEE 802.11 WLAN band (5.725–5.825) GHz application. In simulation it shows a -10 dB impedance bandwidth (5.7–5.86) GHz and the 3 dB AR bandwidth (5.68–5.84) GHz. The mutual coupling is better than -13 dB at the edge frequencies and covers the return loss bandwidth in simulation and measurement as well. The isolation at the center frequency is better than -25 dB in simulation and -35 dB in measurement. The measured impedance bandwidth and 3 dB AR bandwidth are (5.67–5.85) GHz and (5.69–5.84) GHz respectively. This covers the target 5.725–5.825 GHz band for WLAN applications. It can be observed that the antenna operates well as predicted from the simulation. The slight mismatch between the simulation and measurement results may be attributed to factors which can't be counted in the simulation such as simulation environment and tolerance in fabrication.

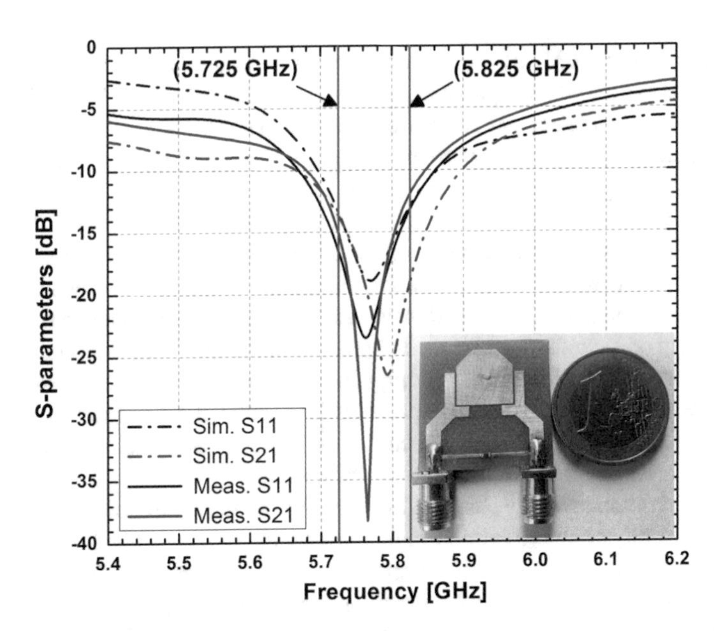

Fig. 6.3 Simulated and measured scattering parameters [188] [J. Malik, A. Patnaik and M.V. Kartikeyan, "Novel Printed MIMO Antenna with Dual-Sense Circular Polarization Diversity", *Proc. of URSI-RCRS 2015, India.* Reproduced courtesy of the URSI-RCRS, Belgium]

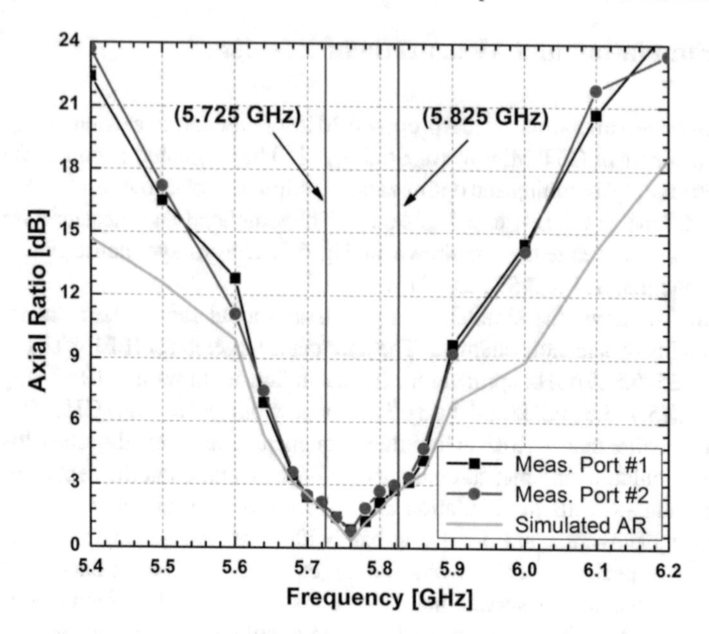

Fig. 6.4 Simulated and measured axial ratio [188] [J. Malik, A. Patnaik and M.V. Kartikeyan, "Novel Printed MIMO Antenna with Dual-Sense Circular Polarization Diversity", *Proc. of URSI-RCRS 2015, India*. Reproduced courtesy of the URSI-RCRS, Belgium]

The axial ratio (AR) and CP performance of a circularly polarized antenna are highly sensitive to its design parameters. In order to achieve a good CP performance, tuning of various design parameters e.g. relative lengths of the arms of L-shape feed-line, chamfer radius of patch, are done intuitively and parametrically. Putting a ground shorting pin at the center of the patch also minimizes mode coupling, suppresses undesirable modes and field neutralization which enhance CP behavior [126]. Due to the symmetrical nature of the antenna structure, the axial ratio for both ports is similar. The simulated AR magnitude against frequency is shown in Fig. 6.4. The AR measurement is performed inside anechoic chamber at discrete frequencies with the help of a standard gain horn antenna and is shown in Fig. 6.4. The measured AR is observed to be little higher compared to the simulated. The measured 3 dB AR bandwidth is also little less compared to the simulated. For a single sense of CP operation, LHCP or RHCP radiation can be easily achieved by selecting one port as the feed port while other port matched terminated.

For polarization diversity purpose, the coupling between the excitation ports, with simulated co-polar and cross-polar radiation patterns are considered. In Fig. 6.3 the antenna is shown to provide good impedance matching and high in-band port-to-port isolation. The simulated 3D radiation patterns for both ports have been shown in Fig. 6.5. It is confirmed that for port-1 right handed circularly polarized waves are radiated in the +Z-axis direction, whereas for port-2, it is left handed circular polarized in the broadside direction. The cross-polarization pattern is also shown in

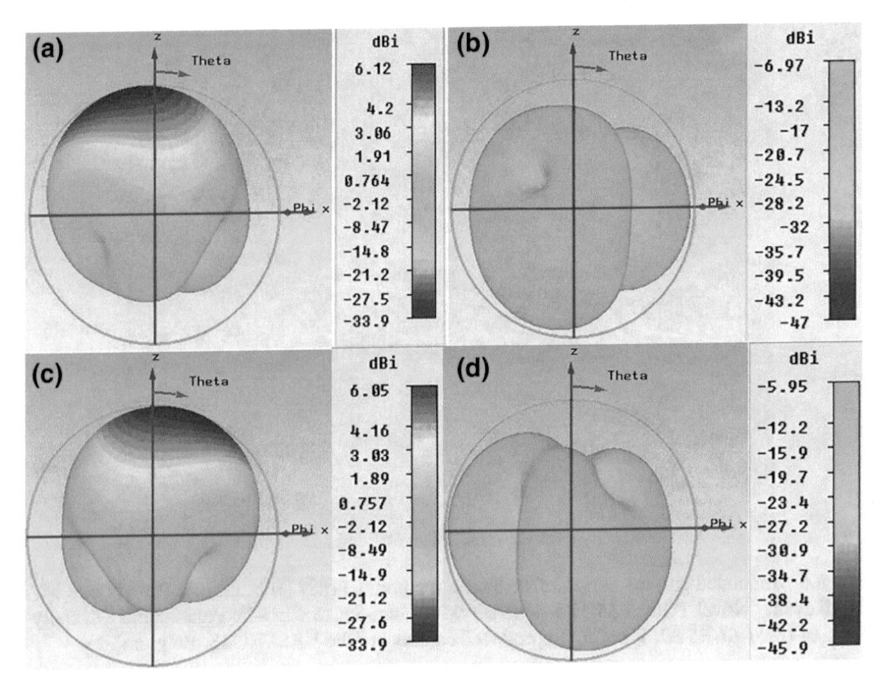

Fig. 6.5 Simulated 3D radiation pattern at 5.8 GHz for port-1 [A] RHCP [B] LHCP; for port-2 [C] LHCP [D] RHCP [188] [J. Malik, A. Patnaik and M.V. Kartikeyan, "Novel Printed MIMO Antenna with Dual-Sense Circular Polarization Diversity", *Proc. of URSI-RCRS 2015, India*. Reproduced courtesy of the URSI-RCRS, Belgium]

Fig. 6.5 for both ports. It can be seen that the cross-polarization level is very less, which ensures good CP operation as desired. The simulated 3 dB AR beamwidth is approximately 100^0 for both polarizations. The wide CP beamwidth ensures good signal reception fidelity. Figures 6.6 and 6.7 shows the simulated surface current distribution on the patch for both ports at different excitation phase. The cyclic rotation of the surface current vector confirms CP behavior of the antenna. It can be observed that the antenna operates with both sense of CP for their intended feeding line.

The radiation pattern for the proposed MIMO antenna is measured inside anechoic chamber. While measuring pattern and gain one port is excited, the other port is matched terminated. The normalized co-polar radiation pattern for both ports is shown in Fig. 6.8. It can be observed that, the XZ-plane patterns are little tilted symmetrically about vertical axis for both ports due to the physical orientation of the feeding lines to the patch.

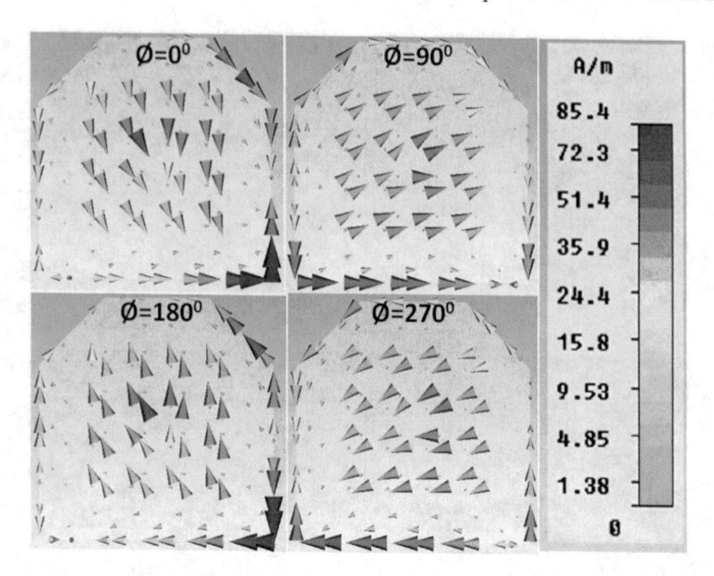

Fig. 6.6 Simulated surface current distribution for port-1 [188] [J. Malik, A. Patnaik and M.V. Kartikeyan, "Novel Printed MIMO Antenna with Dual-Sense Circular Polarization Diversity", *Proc. of URSI-RCRS 2015, India*. Reproduced courtesy of the URSI-RCRS, Belgium]

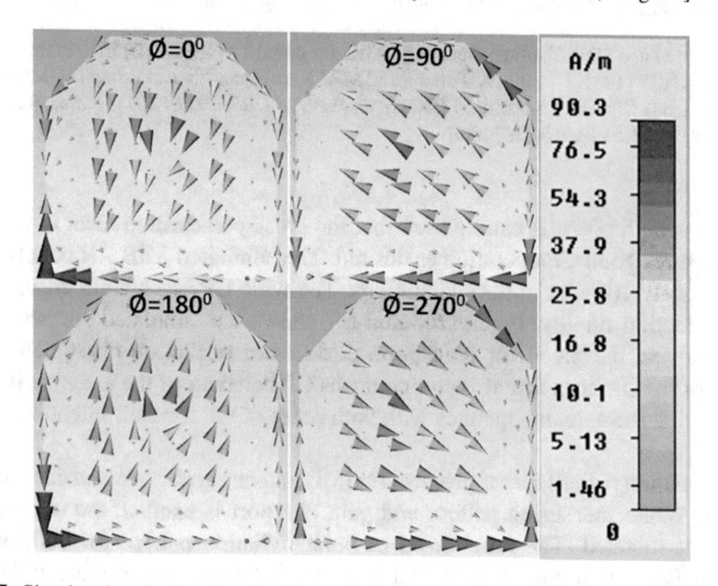

Fig. 6.7 Simulated surface current distribution for port-2 [188] [J. Malik, A. Patnaik and M.V. Kartikeyan, "Novel Printed MIMO Antenna with Dual-Sense Circular Polarization Diversity", *Proc. of URSI-RCRS 2015, India*. Reproduced courtesy of the URSI-RCRS, Belgium]

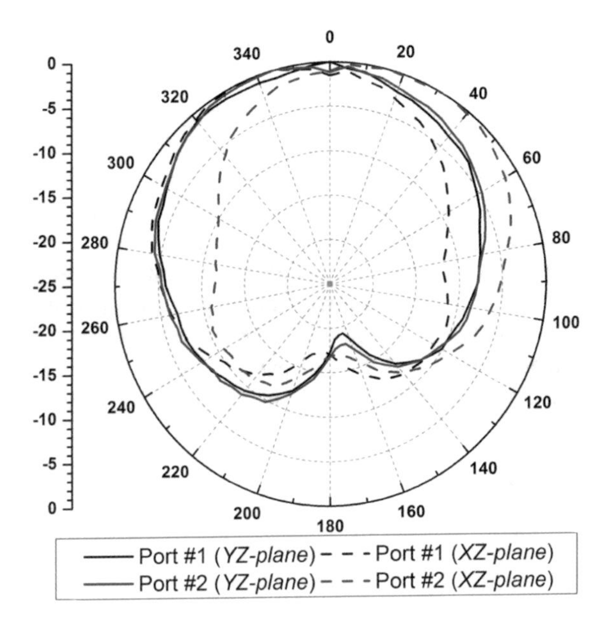

Fig. 6.8 Measured radiation pattern for both ports [188] [J. Malik, A. Patnaik and M.V. Kartikeyan, "Novel Printed MIMO Antenna with Dual-Sense Circular Polarization Diversity", *Proc. of URSI-RCRS 2015, India*. Reproduced courtesy of the URSI-RCRS, Belgium]

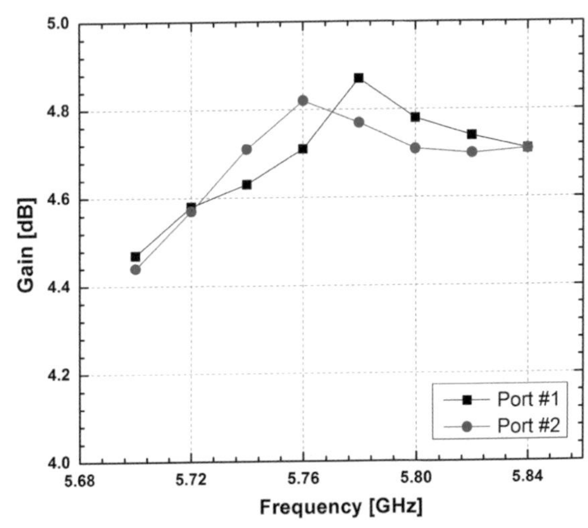

Fig. 6.9 Measured co-polar gain inside anechoic chamber [188] [J. Malik, A. Patnaik and M.V. Kartikeyan, "Novel Printed MIMO Antenna with Dual-Sense Circular Polarization Diversity", *Proc. of URSI-RCRS 2015, India*. Reproduced courtesy of the URSI-RCRS, Belgium]

The measured gain of the MIMO antenna system for both ports is shown in Fig. 6.9. Due to symmetrical nature of the antenna the measured gain for both the polarization shows close similarity.

6.4 Concluding Remarks

A compact MIMO antenna with orthogonal circular polarization diversity have been discussed in this chapter. A L-type feed network have been used to excite circular polarization in a rectangular patch. A decoupling network has been integrated to the MIMO system to improve port-to-port isolation. In measurement the proposed MIMO system shows simultaneous good impedance matching and good 3 dB AR bandwidth. It can be observed that the antenna simultaneously supports both sense of circular polarization with a high port-to-port isolation. The envelope correlation coefficient (ECC) computed from measured scattering parameters and simulated 3D radiation pattern is well below 0.06. This ensures good diversity performance for MIMO applications with circular polarization diversity.

Chapter 7
Conclusion and Future Scope

7.1 Contribution of the Book

In the present work, an attempt has been made to realize compact antenna solutions for high data rate communications systems. The entire work done in the present book is broadly categorized in two parts, i.e. (1) design and analysis of compact UWB antennas for high data rate communication, and (2) design and analysis of compact MIMO antennas for space constrained scenarios.

An introduction of the entire work followed by the motivation behind it and the research objective is presented in Chap. 1 of this book. It also includes scope of the overall work.

Preliminaries and pertinent literature review of printed antennas (particularly UWB and MIMO antennas) for high-speed communication systems are given in Chap. 2.

Chapter 3 deals with time domain analysis of ultra-wideband antennas with notch techniques. In the first section, UWB antenna with single notch has been discussed. In second section, UWB antenna with dual band-notch characteristics has been discussed. In these, the realized notch is static type notch which can't be altered in real time. Sometimes, it is required that the notch to be tuned to reject a particular target frequency. The third section covers UWB antenna with tunable band-notch characteristics.

To mitigate interference with existing narrow band communication systems, UWB antenna with various band-notch techniques has been explored by various researchers. It has been observed that the frequency domain behavior of the UWB antenna remain same for all these band-notch techniques implemented on it. However, time-domain performance of these antennas behaves differently for different band-notch techniques. A time-domain comparison of band-notch techniques has been performed in Chap. 4. A technique is also proposed and proved effective to maintain pulse shape compared to other techniques.

In Chap. 5, compact MIMO antenna solution utilizing pattern diversity has been discussed for space constrained mobile terminals. The first MIMO antenna is designed to have omnidirectional pattern diversity in orthogonal planes. The second MIMO antenna is designed with a combined pattern and polarization diversity.

J. Malik et al., *Compact Antennas for High Data Rate Communication*,
Springer Topics in Signal Processing 14, DOI 10.1007/978-3-319-63175-2_7

In an environment where multiple scattering elements and de-polarizing elements are common, use of a circularly polarized antenna gives excellent performance over linearly polarized antenna. Furthermore, using a circular polarization diversity MIMO antenna, the channel capacity can be increased. Novel antenna solution for MIMO system utilizing circular polarization diversity has been designed and analyzed for a compact terminal in Chap. 6.

In summary, the book contributes towards the development of compact antennas for high data rate communications. Wideband antennas for pulse transmission and compact MIMO antennas with diversity have been designed, analyzed and presented in the book.

We understand that the work done in this book will certainly help the antenna community in solving many issues as far as antenna requirements are there for high-speed communications.

7.2 Future Scope

UWB and MIMO antennas are promising solutions for future high-speed communications. To meet the requirement, compact antenna solutions for successful implementation of these technologies in hand held portable mobile devices is always an open challenge to antenna design community. The following aspects may suitably be explored as the future scope of the present work:

1. Future scope of the work includes designing more compact antenna for the possible UWB and MIMO applications. Methods like use of metamaterials and fractal geometries for size reduction of the designed antennas may be explored.
2. In the present work, antenna design aspects for UWB and MIMO are explored and novel compact antenna designs are presented. However communication system performance with these antennas deployed at transmitter or/and receiver side is not evaluated. A test bed for real-time testing of these antennas can be designed.
3. The propagation environment considered in this work for evaluating correlation in MIMO system is quite simplified and approximate. Real time testing using a MIMO testbed can be done. This would incorporate the effects of actual propagating environment. Verification of increase in channel capacity of a MIMO system over a SISO system can be done. Testing similar designs for multiband operations and their analysis for use in MIMO systems may be carried out.
4. Since the capacity increases linearly with the number of antennas used, this demands for MIMO antennas with large number of radiators. Again coupling is a severe problem that limits the performance of MIMO system. So, novel solutions are required for successful implementation of massive MIMO.

5. The demand for wearable antennas has put challenges on antenna design community. Most of the antennas presented in this work are appropriate for use in portable devices. So, the effects of these antennas on the human body can be studied.
6. Most of the MIMO technologies are implemented on narrow-band services to avoid potential interferences with nearby communication channels. Since, UWB enjoys a wide bandwidth and co-exists with other services, integration of these two technologies may be investigated for high-speed communication. Investigations can be made to realize compact ultra-wideband MIMO antennas having a very wide impedance bandwidth and high in-band isolation between radiators.

References

1. J.G. Proakis, *Digital Communications* (McGraw-Hill, New York, 1989)
2. C.E. Shannon, A mathematical theory of communication. Bell Syst. Techn. J. **27**(379–423), 623–656 (1948)
3. Federal Communication Commission, First report and order, revision of part 15 of the commission's rules regarding ultra-wideband transmission (2002)
4. R.J. Fontana, Recent system applications of short-pulse ultra-wideband (UWB) technology. IEEE Trans. Microwav. Theory Tech. **52**(9), 2087–2104 (2004)
5. L. Yang, G. Giannakis, Ultra-wideband communications: an idea whose time has come. IEEE Signal Process. Mag. **21**(6), 26–54 (2004)
6. K.P. Ray, Design aspects of printed monopole antennas for ultra-wide band applications. Int. J. Antennas Propag. **2008**(713858) (2008)
7. M. Chiani, A. Giorgetti, Coexistence between UWB and narrow-band wireless communication systems. Proc. IEEE **97**(2), 231–254 (2009)
8. A. Paulraj, D. Gore, R.U. Nabar, H. Bolcskei, An overview of MIMO communications-a key to gigabit wireless. Proc. IEEE **92**(2), 198–218 (2004)
9. J. Winters, On the capacity of radio communication system with diversity in a Rayleigh fading environment. IEEE J. Sel. Areas Commun. **5**(5), 871–878 (1987)
10. G. Foschini, M. Gans, On limits of wireless communications in a fading environment when using multiple antennas. Wirel. Pers. Commun. **6**, 311–335 (1998)
11. I. Telatar, Capacity of multi-antenna Gaussian channels. Eur. Trans. Telecommun. **10**(6), 585–595 (1999)
12. M. Jensen, J. Wallace, A review of antennas and propagation for MIMO wireless communications. IEEE Trans. Antennas Propag. **52**(11), 2810–2824 (2004)
13. C.L. Bennett, G.F. Ross, Time-domain electromagnetics and its applications. Proc. IEEE **66**(3), 299–318 (1978)
14. J. Liang, Antenna study and design for ultra-wideband communication, Ph.D. thesis, Queen Mary, University of London, United Kingdom (2006)
15. K. Siwiak, D. McKeown, *Ultra-wideband radio technology* (Wiley, New Jersey, 2004)
16. G.R. Aiello, G.D. Rogerson, Ultra-wideband wireless systems. IEEE Microw. Mag. **4**(2), 36–47 (2003)
17. V. Rumsey, Frequency independent antennas. IRE Natl. Conv. **5**, 114–118 (1957)
18. J. Dyson, The unidirectional equiangular spiral antenna. IRE Trans. Antennas Propag. **7**, 329–334 (1959)
19. G. Ross, A time domain criterion for the design of wideband radiating. IEEE Trans. Antennas Propag. **16**, 355–356 (1968)

J. Malik et al., *Compact Antennas for High Data Rate Communication*,
Springer Topics in Signal Processing 14, DOI 10.1007/978-3-319-63175-2

20. G. Ross, Transmission and reception system for generating and receiving base-band duration pulse signals for short base-band pulse communication system, U.S. Patent 3728632 (1973)
21. C. Fowler, J. Entzminger, J. Corum, Assessment of ultra-wideband (UWB) technology. IEEE Aerosp. Electron. Syst. Mag. **5**(11), 45–49 (1990)
22. F. Nekoogar, F. Dowla, *Ultra-wideband radio frequency identification systems* (Springer, Heidelberg, 2011). ISBN 978-1-4419-9700-5
23. R.C. Qiu, H. Liu, X. Shen, Ultra-wideband for multiple access communications. IEEE Commun. Mag. **43**(2), 80–87 (2005)
24. W. Siriwongpairat, Cross-layer design for multi-antenna ultra-wideband, Ph.D. thesis, University of Maryland (2005)
25. Ultra-wideband (UWB) technology enabling high-speed wireless personal area networks, Intel White Paper (2004)
26. Z.N. Chen, X.H. Wu, H.F. Li, N. Yang, M.Y. Chia, Considerations for source pulses and antennas in UWB radio systems. IEEE Trans. Antennas Propag. **52**(7), 1739–1748 (2004)
27. I. Oppermann, M. Hamalainen, J. Iinatti, *UWB theory and applications* (Wiley, New Jersey, 2004)
28. D.H. Kwon, Effect of antenna gain and group delay variations on pulse-preserving capabilities of ultrawideband antennas. IEEE Trans. Antennas Propag. **54**(8), 2208–2215 (2006)
29. D. Lamensdorf, L. Susman, Baseband-pulse-antenna techniques. IEEE Antennas Propag. Mag. **36**(1), 20–30 (1994)
30. G. Quintero, J.-F. Zurcher, A.K. Skriverviky, System fidelity factor: a new method for comparing UWB antennas. IEEE Trans. Antennas Propag. **59**(7), 2502–2512 (2011)
31. W. Wiesbeck, G. Adamiuk, C. Sturm, Basic properties and design principles of UWB antennas. Proc. IEEE **97**(2), 372–385 (2009)
32. C. Chiau, Study of the diversity antenna array for the MIMO wireless communication systems, Ph.D. thesis, Department of Electronic Engineering, Queen Mary, University of London, United Kingdom (2006)
33. S. Sanayei, A. Nosratinia, Antenna selection in MIMO systems. IEEE Commun. Mag. **42**(10), 68–73 (2004)
34. L. Dong, H. Ling, R. Heath, Multiple input multiple output wireless communication systems using antenna pattern diversity. *IEEE Global Telecommunications Conference (GLOBECOM)*, vol. 1, pp. 997–1001, 17–21 (2002)
35. T. Svantesson, Correlation and channel capacity of MIMO systems employing multimode antennas. IEEE Trans. Veh. Technol. **51**(6), 1304–1312 (2002)
36. C. Waldschmidt, C. Kuhnert, T. Fugen, W. Wiesbeck, Measurements and simulations of compact MIMO-systems based on polarization diversity. IEEE Top. Confer. Wirel. Commun. Technol. 284–285 (2003)
37. M. Nabil Srifi, O. El Mrabet, F. Falcone, M. Sorolla, M. Essaaidi, A novel compact printed circular antenna for very ultra-wideband applications. Microw. Opt. Technol. Lett. **51**(4), 1130–1133 (2009)
38. O. El Mrabet, M. Aznabet, F. Falcone, H. Rmili, J.M. Floc'h, M. Drissi, M. Essaaidi, A compact split ring resonator antenna for wireless communication systems. Progr. Electromag. Res. Lett. **36**, 201–207 (2013)
39. M.M.A. Abaga Abessolo, M. Aznabet, A. Alilouche, O. El Mrabet, M. Essaaidi, M. Beruete, M. Navarro-Ca, F. Falcone, N. Aknin, M. El Moussaoui, M. Sorolla, *Planar Horn Antenna: Application of Periodic Stacked Subwavelength Hole Arrays with Metamaterials Properties* (Mediterranean Microwaves Symposium, Tangier, Morocco, 2009)
40. A.K. Arya, R.S. Aziz, S.O. Park, Planar ultra-wideband printed wide-slot antenna using fork-like tuning stub. Electron. Lett. **51**(7), 550–551 (2015)
41. K.-H. Kim, S.O. Park, Analysis of the small band-rejected antenna with the parasitic strip for UWB. IEEE Trans. Antenna Propag. **54**(6), 1688–1692 (2006)
42. Y.J. Cho, K.H. Kim, D.H. Choi, S.S. Lee, S.O. Park, A miniature UWB planar monopole antenna with 5-GHz band-rejection filter and the time-domain characteristics. IEEE Trans. Antenna Propag. **54**(5), 1453–1460 (2006)

43. K.H. Kim, Y.J. Cho, S.H. Hwang, S.O. Park, Band-notched UWB planar monopole antenna with two parasitic patches. Electron. Lett. **41**(14), 786–788 (2005)
44. Y. Choi, Y. Kim, H. Hoang, F. Bien, A 3.1-4.8 GHz IR-UWB all-digital pulse generator with variable channel selection in 0.13-μm CMOS Technology. IEEE Trans. Circuits Syst. II. **59**(5), 282–286 (2012)
45. K.M. Na, Y.H. Choi, F. Bien, Impulse-radio based ultra-wideband (IR-UWB) transceiver ICs in CMOS for medical implantable devices, in *IEEE International Microwave Symposium*, Seattle, WA, U.S.A., pp. 2–7 (2013)
46. K. Na, H. Jang, H. Ma, Y. Choi, F. Bien, A 200-Mbps data rate 3.1-4.8 GHz IR-UWB all-digital pulse generator with DB-BPSK modulation. IEEE Trans. Circuits Syst. II Expr. Briefs **62**(12), 1184–1188 (2015)
47. T.S. Teeslink, D. Torres, J. Ebel, N. Seplveda, D.E. Anagnostou, Reconfigurable bow tie antenna using metal-insulator transition in vanadium dioxide. IEEE Antennas Wirel. Prop. Lett. **14**, 1381–1384 (2015)
48. M.A.A. Tarifi, D.E. Anagnostou, A.K. Amert, K.W. Whites, The puck antenna: a compact design with wideband, high-gain operation. IEEE Trans. Antennas Propag. **63**(4), 1868–1873 (2015)
49. B.D. Braaten, S. Roy, I. Irfanullah, D.E. Anagnostou, Self-adapting flexible antenna arrays for changing conformal surface applications. IEEE Trans. Antennas Propag. **62**(4), 1880–1887 (2014)
50. C. Sun, A. Hirata, T. Ohira, N.C. Karmakar, Fast beam forming of electronically steerable parasitic array radiator antennas: theory and experiment. IEEE Trans. Antennas Propag. **52**(7), 1819–1832 (2004)
51. N.C. Karmakar, Investigations into a cavity-backed circular-patch antenna. IEEE Trans. Antennas Propag. **50**(12), 1706–1715 (2002)
52. I. Balbin, N.C. Karmakar, Phase-encoded chipless RFID transponder for large-scale low-cost applications. IEEE Microw. Wirel. Components Lett. **19**(8), 509–511 (2009)
53. N.C. Karmakar, S.M. Roy, M.S. Ikram, Development of smart antenna for RFID reader, in *IEEE International Conference on RFID*, Las Vegas, NV, pp. 65–73 (2008)
54. N.C. Karmakar, M.E. Bialkowski, Electronically steerable array antennas for mobile satellite communications—a review, in *IEEE International Conference on Phased Array Systems and Technology*, Dana Point, CA, pp. 81–84 (2000)
55. M. Hassan, S. Dey, N.C. Karmakar, *Ka-band complementary reflector backed slot antenna array for soil moisture radiometer, in International Symposium on Antennas and Propagation (ISAP)* (Hobart, TAS, 2015), pp. 1–4
56. R.K. Chaudhary, K.V. Srivastava, A. Biswas, Three-element multi-layer multi-permittivity cylindrical dielectric resonator antenna for wideband applications with omnidirectional radiation pattern and low cross-polarization. Microw. Opt. Technol. Lett. **54**(9), 2011–2016 (2012)
57. R.K. Chaudhary, K.V. Srivastava, A. Biswas, Wideband multi-layer multi-permittivity half-split cylindrical dielectric resonator antenna. Microw. Opt. Technol. Lett. **54**(11), 2587–2590 (2012)
58. R.K. Chaudhary, K.V. Srivastava, A. Biswas, A practical approach: design of wideband cylindrical dielectric resonator antenna with permittivity variation in axial direction and its fabrication using microwave laminates. Microw. Opt. Technol. Lett. **55**(10), 2282–2188 (2013)
59. K. Krishnamoorthy, B. Majumder, J. Mukherjee, K.P. Ray, Low profile pattern diversity antenna using quarter-mode substrate integrated waveguide. Progr. Electromag. Res. Lett. **55**, 105–111 (2015)
60. G.S. Reddy, A. Kamma, S. Kharche, J. Mukherjee, S.K. Mishra, Cross-configured directional UWB antennas for multi directional pattern diversity characteristics. IEEE Trans. Antennas Propag. **63**(2), 853–858 (2015)
61. P. Patel, B. Mukherjee, J. Mukherjee, A compact wideband rectangular dielectric resonator antenna using perforations and edge grounding. IEEE Antennas Wirel. Propag. Lett. **14**, 490–493 (2015)

62. H.S. Singh, G.K. Pandey, P.K. Bharti, M.K. Meshram, Design and performance investigation of a low profile MIMO/diversity antenna for WLAN/WiMAX/HIPERLAN applications with high isolation. Int. J. RF Microw. Comput. Aided Eng. **25**(6), 510–521 (2015)
63. P.K. Bharti, G.K. Pandey, H.S. Singh, M.K. Meshram, Multiband shorted monopole antenna for handset applications. Microw. Opt. Technol. Lett. **57**(6), 1459–1466 (2015)
64. H.S. Singh, G.K. Pandey, P.K. Bharti, M.K. Meshram, A compact dual-band diversity antenna for WLAN applications with high isolation. Microw. Opt. Technol. Lett. **57**(4), 906–912 (2015)
65. V. Rajeshkumar, S. Raghavan, Bandwidth enhanced compact fractal antenna for UWB applications with 56 GHz band rejection. Microw. Opt. Technol. Lett. **57**(3) (2014)
66. A. Subbarao, S. Raghavan, Printed planar UWB antenna with rejection of WLAN and WiMAX bands. Microw. Opt. Technol. Lett. **55**(4), 740–744 (2013)
67. A. Subbarao, S. Raghavan, A compact UWB slot antenna with signal rejection in 5–6 GHz band. Microw. Opt. Technol. Lett. **54**(5), 1292–1296 (2012)
68. V.V. Reddy, N.V.S.N. Sarma, Triband circularly polarized koch fractal boundary microstrip Antenna. IEEE Antennas Wirel. Propag. Lett. **13**, 1057–1060 (2014)
69. P.N. Rao, N.V.S.N. Sarma, Compact single feed circularly polarized fractal boundary microstrip antenna, in *10th International Conference on Electromagnetic Interference and Compatibility (INCEMIC)*, Bangalore, pp. 347–350 (2008)
70. V.V. Reddy, N.V.S.N. Sarma, Compact circularly polarized asymmetrical fractal boundary microstrip antenna for wireless applications. IEEE Antennas Wirel. Propag. Lett. **13**, 118–121 (2014)
71. W. Zhang, Y.C. Jiao, B. Yang, Z. Hong, CPW-fed ultra-wideband antenna with 3.5/5.5 GHz dual band-notched characteristics, in *IEEE International Conference on Microwave Technology and Computational Electromagnetics (ICMTCE)*, Beijing, China, pp. 327–330 (2011)
72. I.B. Trad, H. Rmili, J. Floch, H. Zangar, Design of planar mono-band rejected UWB CPW-fed antennas for wireless communications, in *11th Mediterranean Microwave Symposium (MMS)*, Hammamet, pp. 101–105 (2011)
73. S. Nikolaou, A. Amadjikpe, J. Papapolymerou, M.M. Tentzeris, Compact ultra-wideband (UWB) elliptical monopole with potentially reconfigurable band rejection characteristic, in *Asia-Pacific Microwave Conference (APMC)*, Bangkok, Thailand, pp. 1–4 (2007)
74. D.-O. Kim, C.-Y. Kim, CPW-fed ultra-wideband antenna with triple-band notch function. Electron. Lett. **46**, 1246–1248 (2010)
75. C. Deng, L. Li, Q. Gong, D. Zhan, Y. Zou, *Planar printed monopole antennas for ultra-wideband/multi-band wireless systems, in IEEE 4th International Symposium on Microwave, Antenna, Propagation, and EMC Technologies for Wireless Communications (MAPE)* (China, Beijing, 2011)
76. T. Li, H. Zhai, L. Li, C. Liang, Y. Han, Compact UWB antenna with tunable band-notched characteristic based on microstrip open loop resonator. IEEE Antennas Wirel. Propag. Lett. **11**, 1584–1587 (2012)
77. N.C. Chauhan, M.V. Kartikeyan, A. Mittal, *Soft computing methods for microwave and millimeter-wave design problems, Studies in Computational Intelligence*, vol. 392 (Springer, Heidelberg, 2012), pp. 69–82
78. T.-G. Ma, S.-K. Jeng, Planar miniature tapered-slot-fed annular slot antennas for ultrawide-band radios. IEEE Trans. Antennas Propag. **53**, 1194–1202 (2005)
79. E.A. Akbari, M.-N. Azarmanesh, S. Soltam, Design of miniaturized band-notch ultra-wideband monopole-slot antenna by modified half-mode substrate-integrated waveguide. IET Microw. Antennas Propag. **7**, 26–34 (2013)
80. Y.D. Dong, W. Hong, Z.Q. Kuai, C. Yu, Y. Zhang, J.Y. Zhou, J.-X C. Dong, Development of ultrawideband antenna with multiple band-notched characteristics using half mode substrate integrated waveguide cavity technology. IEEE Trans. Antennas Propag. **56**(9), 2894–2902 (2008)
81. S.W. Su, K.L. Wong, F.S. Chang, Compact printed ultra-wideband slot antenna with a band notched operation. Microw. Opt. Technol. Lett. **45**(2), 128–130 (2009)

82. A. Mehdipour, A. Parsa, A.R. Sebak, C.W. Trueman, Miniaturised coplanar waveguide-fed antenna and band-notched design for ultrawideband applications. IET Microw. Antennas Propag. **3**(6), 974–986 (2009)
83. Y. Zhang, W. Hong, C. Yu, Z.Q. Kuai, Y.D. Don, J.Y. Zhou, Planar ultrawideband antennas with multiple notched bands based on etched slots on the patch and/or split ring resonators on the feed line. IEEE Trans. Antennas Propag. **56**(9), 3063–3068 (2008)
84. G.M. Yang, R.H. Jin, G.B. Xiao, C. Vittoria, V.G. Harris, N.X. Sun, Ultrawideband (UWB) antennas with multi-resonant split-ring loops. IEEE Trans. Antennas Propag. **57**(1), 256–260 (2009)
85. X. Li, L. Yang, S.X. Gong, Y.J. Yang, Ultra-wideband monopole antenna with four-band-notched characteristics. Progr. Electromag. Res. Lett. **6**, 27–34 (2009)
86. K. Bahadori, Y.R. Samii, A miniaturized elliptic-card UWB antenna with WLAN band rejection for wireless communications. IEEE Trans. Antennas Propag. **55**(11), 3326–3332 (2007)
87. W.-J. Lui, C.-H. Cheng, H.-B. Zhu, Improved frequency notched ultra-wideband slot antenna using square ring resonator. IEEE Trans. Antennas Propag. **55**, 2445–2450 (2007)
88. M. Yazdi, N. Komjani, Design of a band-notched UWB monopole antenna by means of an EBG structure. IEEE Antennas Wirel. Propag. Lett. **10**, 170–173 (2011)
89. Y.D. Dong, W. Hong, Z.Q. Kuai, J.X. Chen, Analysis of planar ultra-wideband antennas with on-ground slot band-notched structures. IEEE Trans. Antennas Propag. **57**, 1886–1893 (2009)
90. L.-N. Zhang, S.-S. Zhong, X.-L. Liang, C.-Z. Du, Compact omnidirectional band-notch ultra-wideband antenna. Electron. Lett. **45**, 659–660 (2009)
91. K.S. Ryu, A.A. Kishk, UWB antenna with single or dual bandnotches for lower WLAN band and upper WLAN band. IEEE Trans. Antennas Propag. **57**, 3942–3950 (2009)
92. C.-J. Pan, C. Lee, C.-Y. Huang, H.-C. Lin, Band-notched ultra-wideband slot antenna. Microw. Opt. Technol. Lett. **48**, 2444–2446 (2006)
93. J. Malik, M.V. Kartikeyan, Band-notched UWB antenna with raised cosine-tapered ground plane. Microw. Opt. Technol. Lett. **56**(11), 2576–2579 (2014)
94. W.-S. Jeong, S.-Y. Lee, W.-G. Lim, H. Lim, J.-W. Yu, Tunable band-notched ultra wideband (UWB) planar monopole antennas using varactor, in 38th European Microwave Conference (EuMC). Amsterdam, Holland **2731**, 266–268 (2008)
95. J. Malik, P.C. Kalaria, M.V. Kartikeyan, Transient response of dual-band-notched ultra-wideband antenna. Int. J. Microw. Wirel. Technol. **7**(1), 61–67 (2014)
96. D.A. Ketzaki, T.V. Yioultsis, Metamaterial-based design of planar compact MIMO monopoles. IEEE Trans. Antennas Propag. **61**(5), 2758–2766 (2013)
97. S.C. Chen, Y.S. Wang, S.J. Chung, A decoupling technique for increasing the port isolation between two strongly coupled antennas. IEEE Trans. Antennas Propag. **56**(12), 3650–3658 (2008)
98. M.A. Moharram, A.A. Kishk, General decoupling network design between two coupled antennas for MIMO applications. Progr. Electromag. Res. Lett. **37**, 133–142 (2013)
99. A.C.K. Mak, C.R. Rowell, R.D. Murch, Isolation enhancement between two closely packed antennas. IEEE Trans. Antennas Propag. **56**(11), 3411–3419 (2008)
100. J. OuYang, F. Yang, Z.M. Wang, Reducing mutual coupling of closely spaced microstrip MIMO antennas for WLAN application. IEEE Antennas Wirel. Propag. Lett. **10**, 310–313 (2011)
101. Z. Li, Z. Du, M. Takahashi, K. Saito, K. Ito, Reducing mutual coupling of MIMO antennas with parasitic elements for mobile terminals. IEEE Trans. Antennas Propag. **60**(2), 473–481 (2012)
102. L. Sun, W. Huang, B. Sun, Q. Sun, J. Fan, Two-port pattern diversity antenna for 3G and 4G MIMO indoor applications. IEEE Antennas Wirel. Propag. Lett. **13**, 1573–1576 (2014)
103. H.T. Chattha, Y. Huang, S.J. Boyes, X. Zhu, Polarization and pattern diversity-based dual-feed planar inverted-F antenna. IEEE Trans. Antennas Propag. **60**(3), 1532–1539 (2012)
104. C.Y. Chiu, J.B. Yan, R.D. Murch, Compact three-port orthogonally polarized MIMO antennas. IEEE Antennas Wirel. Propag. Lett. **6**, 619–622 (2007)

105. J. Oh, K. Sarabandi, Compact, low profile, common aperture polarization, and pattern diversity antennas. IEEE Trans. Antennas Propag. **62**(2), 569–576 (2014)
106. K. Wei, Z. Zhang, W. Chen, Z. Feng, A novel hybrid-fed patch antenna with pattern diversity. IEEE Antennas Wirel. Propag. Lett. **9**, 562–565 (2010)
107. E. Rajo-Iglesias, O. Quevedo-Teruel, S. Fernndez, Compact multimode patch antennas for MIMO applications. IEEE Antennas Propag. Mag. **50**(2), 197–205 (2008)
108. E. Pancera, J. Timmermann, T. Zwick, W. Wiesbeck, Time domain analysis of band notch UWB antennas 3rd. Eur. Confer. Antennas Propag. 3658–3662 (2009). Berlin
109. S. Licul, J.A.N. Noronha, W.A. Davis, D.G. Sweeney, C.R. Anderson, T.M. Bielawa, A parametric study of time-domain characteristics of possible UWB antenna architectures. IEEE Veh. Technol. Confer. (VTC) **5**(3110–3114), 6–9 (2003)
110. T. Dissanayake, K.P. Esselle, Prediction of the notch frequency of slot loaded printed UWB antennas. IEEE Trans. Antennas Propag. **55**(11), 3320–3325 (2007)
111. E. Pancera, D. Modotto, A. Locatelli, F.M. Pigozzo, C. Angelis, Novel design of UWB antenna with band-notch capability. Eur. Confer. Wirel. Technol. **48–50**, 8–10 (2007)
112. S.-W. Qu, J.-L. Li, Q. Xue, A band-notched ultra-wideband printed monopole antenna. IEEE Antennas Wirel. Propag. Lett. **5**(1), 495–498 (2006)
113. J.R. Kelly, P.S. Hall, P. Gardner, Band-notched UWB antenna incorporating a microstrip open-loop resonator. IEEE Trans. Antennas Propag. **59**(8), 3045–3048 (2011)
114. M. Otmani, L. Talbi, T.A. Denidni, Dual polarization diversity reception modeling for indoor propagation channel. IEEE Top. Confer. Wirel. Commun. Technol. 257–258 (2003)
115. F. Dominguez, S. Tenorio, J. Urbano, Circular polarization benefits in HSDPA and MIMO networks, in *Proceedings of the 4thEuropean Conference on Antennas and Propagation (EuCAP)*, vol. 1, no. 5, pp. 12–16 (2010)
116. C. Deng, Y. Li, Z. Zhang, Z. Feng, A compact broadside/conical circularly polarized antenna for pattern diversity design. IEEE Int. Wirel. Symp. (IWS) **1**(3), 24–26 (2014)
117. J.-H. Han, N.-H. Myung, Novel feed network for circular polarization antenna diversity. IEEE Antennas Wirel. Propag. Lett. **13**, 979–982 (2014)
118. F. Yang, Y.R. Samii, A reconfigurable patch antenna using switchable slots for circular polarization diversity. IEEE Microw. Guid. Wave Lett. **12**(3), 96–98 (2002)
119. Y. Yao, X. Wang, X. Chen, J. Yu, S. Liu, Novel diversity/MIMO PIFA antenna with broadband circular polarization for multimode satellite navigation. IEEE Antennas Wirel. Propag. Lett. **11**, 65–68 (2012)
120. C.A. Balanis, *Antenna Theory: Analysis and Design*, 3rd edn. (Wiley, New Jersey, 2005)
121. A. Ghobadi, M. Dehmollaian, Printed circularly polarized Y-shaped monopole antenna. IEEE Antennas Wirel. Propag. Lett. **11**, 22–25 (2012)
122. C.Y. Chiu, C.H. Cheng, R.D. Murch, C.R. Rowell, Reduction of mutual coupling between closely-packed antenna elements. IEEE Trans. Antennas Propag. **55**(6), 1732–1738 (2007)
123. M. Sonkki, E. Salonen, Low mutual coupling between monopole antennas by using two $\lambda/2$ slots. IEEE Antennas Wirel. Propag. Lett. **9**, 138–141 (2010)
124. M.B. Suwailam, M. Boybay, O. Ramahi, Electromagnetic coupling reduction in high-profile monopole antennas using single-negative magnetic metamaterials for MIMO applications. IEEE Trans. Antennas Propag. **58**(9), 2894–2902 (2010)
125. J. Andersen, H. Rasmussen, Decoupling and de-scattering networks for antennas. IEEE Trans. Antennas Propag. **24**(6), 841–846 (1976)
126. J. Huang, iniaturized UHF microstrip antenna for a Mars mission. IEEE Antennas Propag. Soc. Int. Symp. **4**(486–489), 8–13 (2001)
127. L. Jianxin, C.C. Chiau, X. Chen, C.G. Parini, Study of CPW-fed circular disc monopole antenna for ultra wideband applications. IEEE Trans. Antennas Propag. **53**(11), 3500–3504 (2005)
128. J.I. Kim, Y. Jee, Design of ultra-wideband coplanar waveguide-fed LI-shape planar monopole antennas. IEEE Antennas Wirel. Propag. Lett. **6**, 383–387 (2007)
129. D.-N. Chien, A small monopole antenna for UWB mobile applications with WLAN band rejected, in *3rd International Conference on Communications and Electronics (ICCE)*, pp. 379–383, 11–13 (2010)

130. M. Azarmanesh, S. Soltani, P. Lotfi, Design of an ultra-wideband monopole antenna with WiMAX, C and wireless local area network. IET Microw. Antennas Propag. **5**(6), 728–733 (2011)
131. A. Mehdipour, A. Parsa, A.R. Sebak, C.W. Trueman, Planar bell-shaped antenna fed by a CPW for UWB applications. IEEE Antennas Propag. Soc. Int. Symp. **1–4**, 5–11 (2008)
132. S.-J. Kim, J.-W. Baik, Y.-S. Kim, A CPW-fed UWB monopole antenna with switchable notch-band. IEEE Antennas Propag. Soc. Int. Symp. **4641–4644**, 9–15 (2007)
133. J. Jiao, H.-w. Deng, Y.-j. Zhao, Compact ultra-wideband CPW monopole antenna with dual band-notched, in *8th International Symposium on Antennas, Propagation and EM Theory*, Kunming, pp. 263–266, 2–5 (2008)
134. W. Zeng, J. Zhao, Band-notched UWB printed monopole antenna fed by tapered coplanar waveguide, in *IEEE International Conference on Ultra-Wideband (ICUWB)*, vol. 1, pp. 1–4, 20–23 (2010)
135. V.A. Shameena, S. Jacob, C.K. Aanandan, K. Vasudevan, P. Mohanan, A compact CPW fed serrated UWB antenna, in *International Conference on Communications and Signal Processing (ICCSP)*, pp. 108–111, 10–12 (2011)
136. L. Jianjun, Z. Shunshi, K.P. Esselle, A printed elliptical monopole antenna with modified feeding structure for bandwidth enhancement. IEEE Trans. Antennas Propag. **59**(2), 667–670 (2011)
137. P. Thomas, D.D. Krishna, M. Gopikrishna, U.G. Kalappura, C.K. Aanandan, Compact planar ultra-wideband bevelled monopole for portable UWB systems. Electron. Lett. **47**(20), 1112–1114 (2011)
138. C. Deng, Y.-J. Xie, P. Li, CPW-fed planar printed monopole antenna with impedance bandwidth enhanced. IEEE Antennas Wirel. Propag. Lett. **8**, 1394–1397 (2009)
139. H. Zhang, G. Li, J. Wang, X. Yin, A novel coplanar CPW-fed square printed monopole antenna for UWB applications, in *International Conference on Microwave and Millimeter Wave Technology (ICMMT)*, pp. 352–354, 8–11 (2010)
140. W.-J. Li, Q.-X. Chu, A tapered CPW structure half cut disc UWB antenna for USB applications, in *Asia Pacific Microwave Conference (APMC)*, pp. 778–781, 7–10 (2009)
141. A. Kerkhoff, H. Ling, Design of a planar monopole antenna for use with ultra-wideband (UWB) having a band-notched characteristic. IEEE Antennas Propag. Soc. Int. Symp. **830–833**, 22–27 (2003)
142. X. Wang, Z.F. Yao, Z. Cui, L. Luo, S.X. Zhang, CPW-fed band-notched monopole antenna for UWB applications, in *8th International Symposium on Antennas, Propagation and EM Theory*, pp. 204–206, 2–5 (2008)
143. G. Liu, C. Guo, P. Zheng, A new band notched UWB monopole antenna, in *IEEE Asia-Pacific Conference on Applied Electromagnetics (APACE)*, pp. 1–4, 9–11 (2010)
144. W. Zhang, Y.-C. Jiao, B. Yang, Z. Hong, CPW-fed ultra-wideband antenna with 3.5/5.5 GHz dual band-notched characteristics, in *IEEE International Conference on Microwave Technology and Computational Electromagnetics (ICMTCE)*, pp. 327–330, 22–25 (2011)
145. L. Ji, G. Fu, J. Zhao, Q. Lu, CPW-fed UWB antenna with the design of controllable band notch, in *International Conference on Electronics, Communications and Control*, pp. 788–790, 9–11 (2011)
146. D.T. Nguyen, D.H. Lee, H.-C. Park, Very compact printed triple band-notched UWB antenna with quarter-wavelength slots. IEEE Antennas Wirel. Propag. Lett. **11**, 411–414 (2012)
147. T.D. Nguyen, D.H. Lee, H.C. Park, Design and analysis of compact printed triple band-notched UWB antenna. IEEE Antennas Wirel. Propag. Lett. **10**, 403–406 (2011)
148. R. Ghatak, R. Debnath, D.R. Poddar, R.K. Mishra, S.R.B. Chaudhuri, A CPW fed planar monopole band notched UWB antenna with embedded split ring resonators, in *Proceedings of the IEEE Antennas and Propagation Conference, Loughborough*, pp. 645–647, 16–17 (2009)
149. E.A. Daviu, M.C. Fabres, M.F. Bataller, A.V. Jimenez, Active UWB antenna with tunable band-notched behaviour. Electron. Lett. **43**(18), 959–960 (2007)

150. T. Li, H. Zhai, L. Li, C. Liang, Y. Han, Compact UWB antenna with tunable band-notched characteristic based on microstrip open-loop resonator. IEEE Antennas Wirel. Propag. Lett. **11**, 1584–1587 (2012)
151. P.S. Kildal, K. Rosengren, Correlation and capacity of MIMO systems and mutual coupling, radiation efficiency, and diversity gain of their antennas: simulations and measurements in a reverberation chamber. IEEE Commun. Mag. **42**(12), 104–112 (2004)
152. R.G. Vaughan, J.B. Andersen, Antenna diversity in mobile communications. IEEE Trans. Veh. Technol. **36**(4), 147–172 (1987)
153. P. Mattheijssen, M. Herben, G. Dolmans, L. Leyten, Antenna pattern diversity versus space diversity for use at handhelds. IEEE Trans. Veh. Technol. **53**(4), 1035–1042 (2004)
154. M. LeFevre, M.A. Jensen, M.D. Rice, Indoor measurement of handset dual-antenna diversity performance, in *Proceedings of the 47th IEEE Vehicular Technology Conference*, vol. 3, pp. 1763–1769 (1997)
155. S. Zhang, J. Hiong, S. He, MIMO antenna system of two closely positioned PIFAs with high isolation. Electron. Lett. **45**(15), 771–773 (2009)
156. Y. Gao, C.C. Chiau, X. Chen, C.G. Parini, Modified PIFA and its array for MIMO terminals. IEEE Proc. Microw. Antennas Propag. **152**(4), 25–259 (2005)
157. B. Bhattacharyya, Input resistances of horizontal electric and vertical magnetic dipoles over a homogeneous ground. IEEE Trans. Antennas Propag. **11**(3), 261–266 (1963)
158. Y. Chung, S. Jeon, D. Ahn, J. Choi, T. Itoh, High isolation dual-polarized patch antenna using integrated defected ground structure. IEEE Microw. Wirel. Compon. Lett. **14**(1), 4–6 (2004)
159. K. Kim, K. Ahn, The high isolation dual-band inverted F antenna diversity system with the small N-section resonators on the ground plane. Microw. Opt. Technol. Lett. **49**(3), 731–734 (2007)
160. T. Kokkinos, E. Liakou, A. Feresidis, Decoupling antenna elements of PIFA arrays on handheld devices. Electron. Lett. **44**(25), 1442–1444 (2008)
161. I. Kim, C.W. Jung, Y. Kim, Y.E. Kim, Low-profile wideband MIMO antenna with suppressing mutual coupling between two antenans. Microw. Opt. Technol. Lett. **50**(5), 1336–1339 (2008)
162. D. Sievenpiper, L. Zhang, R. Broas, N. Alexopolous, E. Yablonovitch, High-impedance electromagnetic surfaces with a forbidden frequency band. IEEE Trans. Microw. Theory Techn. **47**(11), 2059–2074 (1999)
163. F. Yang, Y.R. Samii, Microstrip antennas integrated with electromagnetic band-gap (EBG) structures: a low mutual coupling design for array applications. IEEE Trans. Antennas Propag. **51**(10), 2936–2946 (2003)
164. M.F. Abedin, M. Ali, Reducing the mutual-coupling between the elements of a printed dipole array using planar EBG structures, in *Proceedings of the IEEE AP-S International Symposium on Antennas and Propagation*, pp. 598–601 (2005)
165. J. Itoh, N. Michishita, H. Morishita, Mutual coupling reduction between two inverted-F antennas using mushroom-type composite right-/left-handed transmission lines, in *Proceedings of the 3rd European Conference on Antennas and Propagation*, pp. 3575–3579, 23–27 (2009)
166. M. Schuhler, R. Wansch, M.A. Hein, Reduced mutual coupling in a compact antenna array using periodic structures, in *Proceedings of the IEEE Antennas and Propagation Conference*, Loughborough, pp. 93–96 (2008)
167. S. Dossche, S. Blanch, J. Romeu, Three different ways to decorrelate two closely spaced monopoles for MIMO applications, in *Proceedings of the IEEE/ACES International Conference on Wireless Communications and Applied Computational Electromagnetics*, pp. 849–852 (2005)
168. T.-I. Lee, Y. Wang, Mode-based information channels in closely coupled dipole pairs. IEEE Trans. Antennas Propag. **56**(12), 3804–3811 (2008)
169. C. Volmer, J. Weber, R. Stephan, K. Blau, M. Hein, An eigen-analysis of compact antenna arrays and its application to port decoupling. IEEE Trans. Antennas Propag. **56**(2), 360–370 (2008)
170. A. Diallo, C. Luxey, P.L. Thuc, R. Staraj, G. Kossiavas, Study and reduction of the mutual coupling between two mobile phone PIFAs operating in the DCS 1800 and UMTS bands. IEEE Trans. Antennas Propag. **54**(11), 3063–3074 (2006)

171. A. Diallo, C. Luxey, P.L. Thuc, R. Staraj, G. Kossiavas, Enhanced two-antenna structures for universal mobile telecommunications system diversity terminals. IET Microw. Antennas Propag. **2**, 93–101 (2008)
172. X. Ling, R. Li, A novel dual-band MIMO antenna array with low mutual coupling for portable wireless devices. IEEE Antennas Wirel. Propag. Lett. **10**, 1039–1042 (2011)
173. S. Zhang, G.F. Pedersen, Mutual coupling reduction for UWB MIMO antennas with a wide-band neutralization line. IEEE Antennas Wirel. Propag. Lett. **15**, 166–169 (2016)
174. S.W. Su, C.T. Lee, F.S. Chang, Printed MIMO-antenna system using neutralization-line technique for wireless USB-dongle applications. IEEE Trans. Antennas Propag. **60**(2), 456–463 (2012)
175. S. Zhang, Z. Ying, J. Xiong, S. He, Ultrawideband MIMO/diversity antennas with a tree-like structure to enhance wideband isolation. IEEE Antenna Wirel. Propag. Lett. **8**, 1279–1282 (2009)
176. R. Addaci, A. Diallo, C. Luxey, P. Le Thuc, R. Staraj, Dual-Band WLAN diversity antenna system with high port-to-port isolation. IEEE Antennas Wirel. Propag. Lett. **11**, 244–247 (2012)
177. C.C. Hsu, K.H. Lin, H.L. Su, Implementation of broadband isolator using metamaterial-inspired resonators and a T-shaped branch for MIMO antennas. IEEE Trans. Antennas Propag. **59**(10), 3936–3939 (2011)
178. J. Sarrazin, Y. Mahe, S. Avrillon, S. Toutain, Collocated microstrip antennas for MIMO systems with a low mutual coupling using mode confinement. IEEE Trans. Antennas Propag. **58**(2), 589–592 (2010)
179. A. Araghi, G. Dadashzadeh, Oriented design of an antenna for MIMO applications using theory of characteristic modes. IEEE Antennas Wirel. Propag. Lett. **11**, 1040–1043 (2012)
180. J. Sarrazin, Y. Mahe, S. Avrillon, S. Toutain, A new multimode antenna for MIMO systems using a mode frequency convergence concept. IEEE Trans. Antennas Propag. **59**(12), 4481–4489 (2011)
181. J. Malik, M.V. Kartikeyan, Band-notched UWB antenna with raised cosine-tapered ground plane. Microw. Opt. Technol. Lett. **56**(11), 2576–2579 (2014)
182. J. Malik, P.C. Kalaria, M.V. Kartikeyan, Transient response of dual-band-notched ultra-wideband antenna. Int. J. Microw. Wirel. Technol. **7**(1), 61–67 (2014)
183. J. Malik, P.K. Velalam, M.V. Kartikeyan, Continuously tunable band-notched ultra-wideband antenna. Microw. Opt. Technol. Lett. **57**(4), 924–928 (2015)
184. J. Malik, A. Patnaik, M.V. Kartikeyan, Time-domain Performance of Band-notch Techniques in UWB Antenna, in *Proceedings of the IEEE Asia-Pacific Microwave Conference (APMC-2016)* (2016)
185. J. Malik, A. Patnaik, M.V. Kartikeyan, Novel band-notch technique for improved time domain performance of printed UWB antenna, in *Proceedings of the of URSI-RCRS 2015, India*
186. J. Malik, D. Nagpal, M.V. Kartikeyan, MIMO antenna with omnidirectional pattern diversity. Electron. Lett. **52**(2), 102–104 (2016)
187. J. Malik, A. Patnaik, M.V. Kartikeyan, Novel printed MIMO antenna with pattern and polarization diversity. IEEE Antennas Wirel. Propag. Lett. **14**, 739–742 (2015)
188. J. Malik, A. Patnaik, M.V. Kartikeyan, Novel printed MIMO antenna with dual-sense circular polarization diversity, in *Proceedings of URSI-RCRS 2015, India*
189. J. Malik, A. Patnaik, M.V. Kartikeyan, Printed MIMO antenna with circular polarizatrion diversity, Private communication and also Chap. 4, Ph.D. thesis, J. Malik, Department of Electronics and Communication Engineering, Indian institute of Technology Roorkee, Roorkee, India (2016)

Printed in the United States
By Bookmasters